AMTE Monograph Series Volume 6

Scholarly Practices and Inquiry in the Preparation of Mathematics Teachers

Edited by

Denise S. Mewborn
University of Georgia

Hollylynne S. Lee
North Carolina State University

Monograph Series Editor

Marilyn E. Strutchens
Auburn University

Published by
Association of Mathematics Teacher Educators
San Diego State University
c/o Center for Research in Mathematics and Science Education
6475 Alvarado Road, Suite 206
San Diego, CA 92129

www.amte.net

Library of Congress Cataloging-in-Publication Data

Inquiry into mathematics teacher education / edited by Denise S. Mewborn,
Hollylynne S. Lee.
    p. cm. -- (AMTE monograph series ; v. 6)

ISBN 978-1-62396-951-6

1. Mathematics teachers--Training of--United States. 2. Mathematics—Study
and teaching--United States. 3. Student Teachers. I. Mewborn, Denise S. II.
Lee, Hollylynne S.

QA183.3.A1S346 2009
510.71'073—dc22
                       2009045419

The publications of the Association of Mathematics Teacher
Educators present a variety of viewpoints. The views expressed or
implied in this publication, unless otherwise noted, should not be
interpreted as official positions of the Association.

Contents

   Hala N. Ghousseini, University of Michigan

Reys, B. J.
AMTE Monograph 6
Scholarly Practices and Inquiry in the Preparation of Mathematics Teachers
© 2009, pp. vii-ix

Foreword

On behalf of the Association of Mathematics Teacher Educators (AMTE), I am pleased to introduce this important resource for the field of mathematics teacher education. The sixth monograph of AMTE, *Scholarly Practices and Inquiry in the Preparation of Mathematics Teachers,* edited by Denise S. Mewborn and Hollylynne S. Lee, highlights examples of the important scholarship of the mathematics teacher education community.

This monograph, like others produced by AMTE, serves as a forum for mathematics teacher educators to exchange ideas, experiences, resources and detailed accounts of work to improve preservice and inservice teacher preparation. Chapters address important issues such as: designing tasks to emphasis mathematics knowledge for teaching; capitalizing on opportunities for student teaching mentor learning; and learning to lead classroom mathematics discussions.

AMTE is pleased to support the dissemination of knowledge important to the field. While the monograph series has served as an important vehicle, we realize that what is needed is a more frequent and accessible outlet for the knowledge accumulated by the field. As noted by the co-editors in the opening chapter of this monograph,

> ... there are few outlets for manuscripts that report on scholarly practices in mathematics teacher education. Such a journal would be akin to the practitioner journals that exist for mathematics teachers at various levels. At present the only options for these types of manuscripts are this monograph and journals outside the field of teacher education that focus on the scholarship of teaching (e.g., *The Journal of Scholarship of Teaching and Learning*). While it is beneficial to publish work in mathematics teacher education outside the field, doing so makes it more difficult for other mathematics teacher educators to be aware of this work and to develop a coordinated knowledge base. (p. 3)

AMTE is responding to this need by initiating a practitioner-based journal. The journal will contribute to building a professional knowledge base in mathematics teacher education that stems from practitioner knowledge that is not only public, shared, and stored, but also verified and improved over time (Hiebert, Gallimore & Stigler, 2002). The AMTE Journal Task Force, chaired by Alfinio Flores (AMTE Publications Director), is currently working to conceptualize the journal and gather information to inform long-term planning.

AMTE has also committed to developing a special issue of JMTE focused on equity in mathematics teacher education. The special issue is edited by Marilyn E. Strutchens. It will feature articles that report on research outcomes that will inform the field on how to best address equity issues in the mathematics classroom and other factors that impact equity in teacher education across the continuum from preparation to early career to experienced teacher.

As noted, AMTE dissemination efforts are expanding, due in large part to the consistently high quality of the AMTE monograph series. This present installment, *Scholarly Practices and Inquiry in the Preparation of Mathematics Teachers*, is no exception and further solidifies both the need for and the quality of work in mathematics teacher education.

On behalf of AMTE, I thank those involved in the development of this sixth AMTE monograph including:

Co-Editors
Denise S. Mewborn, *University of Georgia*
Hollylynne S. Lee, *North Carolina State University*

Editorial Panel
Laurie Cavey, *James Madison University*
Rebekah Elliott, *Oregon State University*
Alfinio Flores, *University of Delaware*
Suzanne Harper, *Miami University of Ohio*
Kate Kline, *Western Michigan University*
Johnny Lott, *University of Mississippi*
Jennifer Luebeck, *University of Montana*
Lew Romagnano, *The Metropolitan State College of Denver*
Gideon L. Weinstein, *Western Governors University*

AMTE Monograph Series Editor
Marilyn E. Strutchens, *Auburn University*

Collectively, the authors, editors and editorial panel have prepared a resource that is important for the field.

Barbara J. Reys
AMTE President 2009–2011

Reference

Hiebert, J., Gallimore, R., & Stigler, J. W. 2002. A knowledge base for the teaching profession: What would it look like and how can we get one? *Educational Researcher, 31*(5), 3–15.

Lee, H. S. and Mewborn D. S.
AMTE Monograph 6
Scholarly Practices and Inquiry in the Preparation of Mathematics Teachers
© 2009, pp. 1–6

1

Mathematics Teacher Educators Engaging in Scholarly Practices and Inquiry

Hollylynne S. Lee
North Carolina State University

Denise S. Mewborn
University of Georgia

Scholarship is foundational to all fields of study in order to push the boundaries of our professional knowledge. In many fields, those who engage in the practice of the discipline and those who produce scholarship in the discipline are different groups of people. However, in mathematics teacher education we have a rare opportunity to play both roles. Our practice produces researchable questions, and the results of our scholarship feed back into our practice, thus producing different questions in a repeating cycle. When we harness the combined efforts of multiple scholars in the field, we can begin to gain a critical mass of information that moves the field, and not just our individual practices, forward. Mathematics teacher education is a field in which the scholarship of teaching can be played out in its truest sense.

The scholarship of teaching is a relatively new conception in higher education, stemming largely from the work of Boyer (1990), and there has been much discussion and debate about the scope, intent, and purpose of this type of work. As illustrated by Reed and Mathews (2008), mathematics teacher educators practicing within institutions of higher education are engaging in conversations and debates within their institutions about what are considered "scholarly" contributions to the field of mathematics education and mathematics teacher education. Richlin (2001)

1

helped to distinguish between scholarly teaching practices and scholarship of teaching by offering that

> the purpose of scholarly teaching is to impact the activity of teaching and the resulting learning, whereas the scholarship of teaching results in a formal, peer-reviewed communication in the appropriate media or venue, which then becomes part of the knowledge base of teaching and learning in higher education. (p. 58)

The scholarship of mathematics teacher education focuses on the issues and processes involved in the education of mathematics teachers at all levels. Examples of contemporary issues of interest include:

- the structure of teacher education programs (e.g., the number, type, focus, and sequencing of courses) and professional development programs and their impact on teachers' learning (and subsequent students' learning);
- recruitment and retention of mathematics teachers; or
- the role of classroom based mentors and university supervisors in field-based teaching experiences.

Processes of interest often include how teachers:

- develop mathematical knowledge for teaching;
- learn to examine and make sense of students' mathematical work;
- develop strategies for classroom discourse; or
- learn to create and implement mathematical tasks with a high cognitive demand.

Research questions focused on these issues and processes have emerged from our practices as mathematics teacher educators, and insights into these questions have informed our practices and are beginning to help shape the professional knowledge base of our field (see for example Sowder, 2007).

Monograph IV: Contributions to the Professional Knowledge Base

In the prior AMTE monograph, Arbaugh and Taylor (2008) advocated that the field of mathematics teacher education needed to look toward building a research program that could help coordinate the knowledge base in our field. They discussed the difference between practical knowledge (based on experiences and reflections) and professional knowledge (based on empirical research) and presented a framework adapted from Borko (2004) that could be used to help frame the research in mathematics teacher education and establish "a deeper, more connected professional knowledge base" (Arbaugh & Taylor, 2008, p. 5).

This current monograph contributes to that call by highlighting examples of the scholarship of mathematics teacher education. Some of this scholarship takes the form of reports of those who have engaged in *scholarly practices* in mathematics teacher education—practices adapted from empirical studies of the teaching and learning of mathematics and the preparation of mathematics teachers. The manuscripts that report on these scholarly practices provide evidence of the purposeful synthesis and application of professional knowledge in experiences designed for mathematics teachers (preservice or inservice) and critical reflection on the impact of these experiences on teachers' learning.

Other manuscripts in this monograph illustrate *scholarly inquiry* into issues and practices through systematic data collection and analysis that yields theoretically grounded and empirically based findings. Some of these findings are most important at the local level as they can inform the design of learning opportunities for teachers by the researchers as well as by others. However, some of the findings, when considered along with other research in mathematics teacher education, can contribute on a more global level to the professional knowledge base in our field.

Looking to the Future: Outlets for Scholarly Inquiry in Mathematics Teacher Education

One of the challenges faced by our field is the lack of publication outlets for manuscripts based on the scholarship of teaching. Manuscripts that take the scholarly inquiry approach noted above can be submitted to the *Journal of Mathematics Teacher Education* or a number of other journals in the field that, while not exclusively devoted to issues of mathematics teacher education, publish research in the broader fields of teacher education or mathematics education. However, there are few outlets for manuscripts that report on scholarly practices in mathematics teacher education. Such a journal would be akin to the practitioner journals that exist for mathematics teachers at various levels. At present the only options for these types of manuscripts are this monograph and journals outside the field of teacher education that focus on the scholarship of teaching (e.g., *The Journal of Scholarship of Teaching and Learning*). While it is beneficial to publish work in mathematics teacher education outside the field, doing so makes it more difficult for other mathematics teacher educators to be aware of this work and to develop a coordinated knowledge base.

Given that the investigation of scholarly practices often spurs subsequent scholarly inquiry, it seems important to have an outlet in our field for reports of scholarly practices. As more doctoral programs place an emphasis on the preparation of mathematics teacher educators, both as practitioners and scholars (as recommended by Association of Mathematics Teacher Educators, 2002 and Wilson & Franke, 2008), and as the field continues to grow and develop, the need for such an outlet will become more acute. With the upcoming hiatus of the AMTE monograph, it is imperative that the field seek ways to meet this need. *AMTE* is leading the way with a task force to develop a plan for a practitioner-oriented mathematics teacher education journal.

We are pleased to have had the opportunity to help shape the sixth *AMTE* monograph and hope that readers will be inspired in their own practice and scholarship by what they find in these pages. We thank the authors for sharing their ideas with the field and for working diligently with us on editorial revisions. We appreciate the work of our editorial review panel members who provided valuable critiques and insights into each manuscript reviewed for possible inclusion and thank Marilyn Strutchens as our series editor for all her work in compiling and polishing the chapters to prepare them for publication. A great deal of behind the scenes work goes into crafting an edited volume, and we particularly wish to thank Allyson Hallman and Eric Gold, doctoral students at the University of Georgia for their help during the review process. Allyson continued to work with the manuscripts during the editing process as well, lending a careful eye and thoughtful analysis to each manuscript.

References

Arbaugh, F., & Taylor, M. P. (2008). Inquiring into mathematics teacher education. In F. Arbaugh & M. P. Taylor (Eds.), *Inquiry into mathematics teacher education (AMTE Monograph V)* (pp. 1–9). San Diego, CA: Association of Mathematics Teacher Education.

Association of Mathematics Teacher Educators. (2002). *Principles to guide the design and implementation of doctoral programs in mathematics education* (Task Force Report). San Diego, CA: AMTE. Available online at http://www.amte.net/resources/Doctoral_Studies_position_paper.htm

Borko, H. (2004). Professional development and teacher learning: Mapping the terrain. *Educational Researcher 33(8),* 3–15.

Boyer, E. L. (1990). *Scholarship reconsidered: Priorities of the professoriate.* Princeton, NJ: Carnegie Foundation for the Advancement of Teaching.

Reed, M. K., & Mathews, S. M. (2008). Scholarship for mathematics educators: How does this count for promotion and tenure? In F. Arbaugh & M. P. Taylor (Eds.), *Inquiry into mathematics teacher education (AMTE Monograph V)*

(pp. 157–166). San Diego, CA: Association of Mathematics
Teacher Education.

Richlin, L. (2001). Scholarly teaching and the scholarship of
teaching. *New Directions for Teaching and Learning 86*, 57–
68.

Sowder, J. (2007). The mathematical education and development
of teachers. In F. Lester (Ed.), *Second handbook of research
on mathematics teaching and learning* (pp. 157– 223).
Reston, VA: National Council of Teachers of Mathematics
& Charlotte, NC: Information Age Publishing.

Wilson, P. S., & Franke, M. (2008). Preparing teachers in
mathematics education doctoral programs: Tensions and
strategies. In R. E. Reys & J. A. Dossey (Eds.), *U.S.
doctorates in mathematics education: Developing stewards
of the discipline* (pp. 103–110). Washington, DC: American
Mathematical Society/Mathematical Association of America.

Hollylynne S. Lee is an associate professor and graduate
program coordinator of mathematics education in the department
of Mathematics, Science, and Technology Education at North
Carolina State University. Her scholarly practices and inquiry
engage her in scholarship related to the teaching and learning of
mathematics with technology, with a particular focus on
probability and statistical concepts.

Denise S. Mewborn is professor of mathematics education and
head of the Department of Mathematics and Science Education
at the University of Georgia. Her teaching and research focus on
preservice elementary mathematics teacher education and beliefs
development.

Suzuka, K., Sleep, L., Ball, D. L., Bass, H., Lewis, J. M., and Thames, M. K.
AMTE Monograph 6
Scholarly Practices and Inquiry in the Preparation of Mathematics Teachers
© 2009, pp. 7–23

2

Designing and Using Tasks to Teach Mathematical Knowledge for Teaching[i]

Kara Suzuka, Laurie Sleep, Deborah Loewenberg Ball,
Hyman Bass, Jennifer M. Lewis and Mark Hoover Thames
University of Michigan

Teaching is mathematically demanding work. The requisite knowledge and skills are not necessarily "picked up" on the job nor are they typically learned in college courses or used in other professions. Our research group has been working to better understand the mathematical demands of teachers' work and to find ways to provide teachers with opportunities to develop facility with mathematical knowledge for teaching (MKT). In this paper, we discuss our efforts to design tasks that create such opportunities as well as to identify and study important features of MKT tasks. We also explore some of the challenges of enacting MKT tasks with teachers.

Research over the last two decades clearly shows that teachers make a difference in student achievement (National Mathematics Advisory Panel, 2008). In the whirlwind of problems that plague mathematics education in the U.S.—from weak student achievement, to a repetitive and unfocused curriculum, to persistent racial and income-related gaps in learning opportunities—one important and promising finding is that instruction matters and that teachers are key to effective instruction.

Teachers' skill in making mathematics accessible to and learnable by all students depends on a wide range of resources, including their own mathematical knowledge. Knowing mathematics for teaching can enable sensitive analysis of a student's difficulty, skilled use of diagrams or models, clear explanations, well-posed questions, and strategic selection and use of tasks and examples. However, the mathematical knowledge that supports effective teaching involves more than sheer adeptness with the school curriculum. It requires learning more and different mathematics. Furthermore, such knowledge is not necessarily the product of conventional mathematics study (Begle, 1979; Monk 1994; National Mathematics Advisory Panel, 2008) because such study is not usually aimed at the specialized knowledge of mathematics needed for instruction.

This chapter describes our ongoing efforts to create opportunities for teachers—both practicing and preservice—to develop the mathematical knowledge and skills demanded by the work of teaching. We begin with a brief introduction to our practice-based theory of *mathematical knowledge for teaching* (MKT). We then describe our work to design tasks that provide opportunities to develop MKT, and propose a preliminary list of features that we argue make a task well suited for helping people develop mathematical knowledge for teaching. We follow this with a discussion of some of the challenges of enacting MKT tasks and conclude with ideas for future research in this area.

Mathematical Knowledge for Teaching

A first step in designing opportunities for teachers to learn mathematics is to identify the content and skills needed for their work. Certainly, teachers must be facile with the mathematics of the school curriculum. Consider, for example, an elementary or middle school teacher who is teaching her students how to divide fractions and asks them to do the following calculation: $\frac{5}{6} \div \frac{1}{3}$.

She must obviously be able to compute the answer correctly herself; however, this is far from sufficient. Teaching requires mathematical knowledge and reasoning beyond what students are learning.

As a case in point, suppose one of the teacher's students solves the division problem using the following method: $\frac{5}{6} \div \frac{1}{3} = \frac{10}{12} \div \frac{4}{12} = 10 \div 4 = 2\frac{1}{2}$. Although this is not a standard procedure, the answer is correct. Is this a coincidence, or might this be a mathematically valid method? If the method "works," does it work in general or only with specific numbers? Why can the denominators of like fractions be "ignored" with division (as in the example above), but not with multiplication, addition, or subtraction? Questions such as these pervade the work of teaching. Being able to ask and answer them depends on more than being able to get right answers oneself. We call this *specialized* knowledge of mathematics, because it is needed for the special demands of teaching (Ball, Thames, & Phelps, 2008). Although it might be nice if the average adult were also able to do these sorts of analyses and answer these sorts of questions, for teachers it is essential.

Analyzing student solutions—both correct and incorrect—is just one example of the many *mathematical* tasks of teaching. Teaching also entails explaining mathematical concepts and procedures, making clear, for instance, why equivalent fractions can be generated by multiplying the numerator and denominator by the same number. Teaching requires responding to students' "why" questions, such as why the quotient, 2½ in the calculation above, is a larger number than both the dividend and the divisor. Teaching also requires knowing that "division makes smaller" is a misconception that students are likely to hold from their work with whole numbers. Teachers must select numbers for problems and sequence their use; they must choose representations to demonstrate a procedure or highlight a particular idea. All of these are routine tasks of teaching, and all require mathematical knowledge and reasoning specific to the work that teachers do (Ball & Bass, 2003; Ball, Sleep, Boerst, & Bass, (2009); Ball, Thames, & Phelps, 2008).

Over the past decade, we and our colleagues have worked to identify and articulate the mathematical demands of teaching practice. Using extensive records of classroom interactions, we look in fine detail at the *mathematical* issues that arise (Ball &

Bass, 2003). Our analyses have led to a practice-based theory of *mathematical knowledge for teaching* (MKT)—the mathematical knowledge, skills, and habits of mind entailed by the work of teaching[ii] (Ball, Hill, & Bass, 2005; Ball et al., 2008).

Designing Tasks for Teaching MKT

A better understanding of the ways that teachers know and use mathematics in their work raises important questions for teacher education: What kinds of learning opportunities would help teachers develop such mathematical knowledge, skills, and dispositions? How can we design tasks that create such opportunities for teachers? Teachers in the United States typically learn mathematics in settings that focus exclusively on mathematics. Sometimes, in such contexts, the mathematics is quite general, designed for anyone studying mathematics regardless of how they may or may not use it in their work. There are also many examples of mathematical experiences that are designed to help teachers—elementary teachers in particular—develop a sense of competence with mathematics, begin to see themselves as "mathematical," or simply encounter mathematics as a much broader human endeavor than what they might have known as students. These are worthwhile outcomes (e.g., Schifter & Fosnot, 1993). However, our success as a field in preparing teachers and supporting their ongoing professional learning must extend beyond equipping them with a broader, deeper *personal* understanding of mathematics. We also must help them learn mathematics in ways that enable them to deploy mathematical knowledge and skills in the service of helping *others* learn and do mathematics. Otherwise, as is far too often the case, teachers are left to figure out on their own how to make use of the mathematics they have learned to improve their instruction.

Our efforts to create opportunities for learning mathematical knowledge for teaching have centered on designing tasks that allow people to explore and practice the kind of specialized mathematics used in teaching. To get a better sense of what we mean by such "MKT tasks," we describe an example task below, one that has been used in a number of studies on teacher knowledge (Ball, 1988; Ma, 1999).

An Example MKT Task: 1¾ ÷ ½

Consider the division expression 1¾ ÷ ½. Teachers could be asked to compute the answer to this problem, and certainly, all teachers of mathematics need to be able to do so correctly. However, to focus the task on *MKT*, teachers could also be asked to write a story problem that corresponds to the division. Although some might argue that everyone should understand fractions and the meaning of division well enough to generate a sensible story problem context or that this would be a good task for children, the task of writing a story problem provides an opportunity to develop the sorts of knowledge and skills *teachers* routinely draw upon for their work. For example, teachers often need to create problems that accurately represent a mathematical situation to illustrate an idea to the class or, alternatively, to provide students with mathematics problems that connect to their lives.

Another MKT task focused on this same division problem asks teachers to analyze the following *incorrect* story problem:

I have two pizzas. My friend eats one quarter of one of the pizzas and then I split what's left evenly between two of my other friends, Dan and Heather, and each person gets three and a half pieces of pizzas.

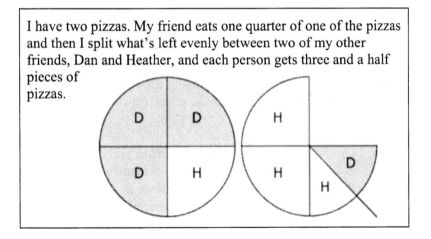

In this task, teachers must first figure out what is wrong with the story despite its seemingly correct answer of 3½. Once this is resolved, teachers are then asked to write a story problem that correctly represents the division. This error-analysis task is squarely situated in an essential part of teaching—that is, thinking about someone else's mathematical solution. In fact, attending to others' thinking is a distinguishing characteristic of the work of *teaching* mathematics as opposed to just doing mathematics for oneself. Teachers routinely study and respond to others' approaches and solutions rather than merely constructing and revising their own. At the same time, the work entailed in this task—analyzing representations and their equivalence, examining the mapping between a model and the concept or procedure it represents, and figuring out where a solution went wrong—is fundamentally *mathematical* in nature. It is on this special interweaving of the practices of teaching and mathematics that we focus our tasks.

Features of MKT Tasks

As illustrated above, there are certain elements of a task—such as providing opportunities to consider the thinking of others—that shift its focus from developing personal knowledge of mathematics to developing MKT. Having a better sense of what makes a task an "MKT task" would enable the field to design more targeted materials for mathematics teacher education and professional development. Although our own thinking continues to evolve, we propose the following as a preliminary list of features that make a task well suited for developing MKT:

- Creates opportunities to unpack, make explicit, and develop a flexible understanding of mathematical ideas that are central to the school curriculum
- Provokes a stumble due to a superficial "understanding" of an idea
- Opens opportunities to build connections among mathematical ideas

- Lends itself to alternative/multiple representations and solution methods
- Provides opportunities to engage in mathematical practices central to teaching (e.g., explaining, representing, using mathematical language, analyzing equivalences, proving, analyzing proofs, posing questions)

We elaborate and illustrate these features below.

First, tasks that develop mathematical knowledge for teaching create opportunities for teachers to unpack or make explicit mathematical ideas that are central to the school curriculum. For example, the 1¾ ÷ ½ error-analysis task engages solvers with core ideas about fractions, such as the importance of attending to the referent unit or whole.

However, the 1¾ ÷ ½ task is designed to do more than develop understandings of fractions. The use of fractions aims to *provoke a stumble* due to a superficial understanding of division. This is a second feature of MKT tasks: They surface a common error or counterintuitive result[iii] that must be confronted and reconciled. Solving the problem creates an opportunity to reconsider assumptions and revisit fragile or partial understandings. In fact, the particular numbers used in this task contribute significantly to the challenge. If 1¾ and ½ were replaced with two other fractions, the problem would not work quite as well. The divisor of ½ is important because dividing *by* one half is often confused with dividing *in* half, leading to the common error of writing a story that corresponds to division by two. In addition, these two numbers—1¾ and ½—lead people to a wrong answer that seems right. They arrive at a quotient of three and a half *fourths,* and because this "3½" matches the calculated answer they fail to notice that their story problem is incorrect. It is precisely this stumble that creates the opportunity to think more deeply about the meaning of the division.

Another feature of MKT tasks is that they offer opportunities for teachers to build connections among mathematical ideas, thus helping teachers develop a more coherent view of the mathematical landscape. For example, work on the 1¾ ÷ ½ task prompts examination of different meanings of division (e.g., as "measurement" or as "partitioning"). When interpreted as measurement, division determines how many groups of a certain size (i.e., the divisor) "fit" into the dividend: In this case, how many halves fit into one and three fourths? Alternatively, division also can be seen as "partitioning" the dividend into a certain number (i.e., the divisor) of groups, with the quotient representing the size of one group or partition. Interpreting $1\frac{3}{4} \div \frac{1}{2}$ in this way requires partitioning $1\frac{3}{4}$ into one half of a group. Although it is a bit counter-intuitive to talk about a fractional number of groups, the quotient sensibly indicates the size of the group for which $1\frac{3}{4}$ is half—specifically, a group of size $3\frac{1}{2}$. Different interpretations of numbers and operations arise repeatedly throughout the school curriculum—e.g., subtraction as "taking away" or the comparison of difference; multiplication as repeated addition or the area of a rectangle; fractions as a ratio, quotient, or relationship of a part to a whole. MKT tasks provide teachers with opportunities to build these types of foundational understandings and explore recurring themes in mathematical work.

Fourth, it is important that MKT tasks lend themselves to multiple solutions or representations. Teaching involves recognizing the wide range of methods and ideas that students produce as well as the ability to discriminate among those that are and are not mathematically valid. If teachers encounter only their own answers, they will not be adequately equipped for the mathematical demands of their work that routinely involves the analysis of children's answers. When we discuss multiple solutions or representations in our teacher education courses, preservice teachers will sometimes say, "I get so confused when I have to hear other people's answers. I thought I understood it, and then I hear someone else's thinking; and I get completely confused. I really dislike it when we do that." Engaging with others' (sometimes confusing) ideas is, in fact, precisely why we pose this type of task. Although such discomfort with confusion is understandable, avoiding it is not viable in teaching. In teaching, teachers do not simply think about and use their own ideas and methods. Working through confusion to more robust understandings is a crucial part of developing MKT. Therefore, we seek to construct tasks that precipitate multiple solutions and representations—both valid and invalid—and create opportunities to work on them.

Finally, tasks that focus on MKT should provide opportunities for teachers to engage in the actual mathematical practices central to teaching—for example, explaining mathematical concepts, making diagrams, or analyzing the equivalence of two seemingly different solutions. Important to note is that these mathematical tasks are fundamental to teaching across a range of pedagogical approaches. That is, practices such as explaining a mathematical idea are a central component of mathematics instruction, regardless of the curriculum or "teaching style" being used. Whether a teacher explains a mathematical idea to the class or scaffolds a student's effort to do so, the work of teaching mathematics involves understanding and skill with practices central to the discipline.

Enacting MKT Tasks

Well-designed tasks are not, by themselves, adequate for helping teachers develop the mathematical knowledge and skills needed for the work of teaching. How these tasks are *enacted* in teacher education and professional development contexts matters. Stein, Smith, Henningsen, and Silver (2000) have developed a useful framework, known as the "Mathematical Tasks Framework," that conceptualizes the transformation of mathematics tasks as they initially appear in curriculum materials and eventually result in some form of student learning (see Figure 1).

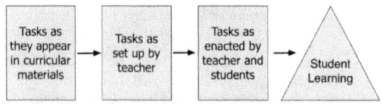

Figure 1: Mathematical Tasks Framework (Stein, Smith, Henningsen, & Silver, 2000).

Tasks written in curriculum materials embody the intentions and designs of their developers. They undergo a first transformation as they are interpreted and set up for use by teachers and are further transformed through the interactions of teachers and students as they engage in the task. Through the back-and-forth interplay of questions, responses, initial attempts, and feedback, tasks are adjusted to meet the perceived needs and desires of individuals and the class as a whole. Finally, mathematical tasks result in some sort of student learning. However, what students learn through their work on a task may differ quite a bit from what was intended by the task's authors or by the teacher who implemented it.

Challenges of Enacting MKT Tasks

The Mathematical Tasks Framework provides a useful lens for considering the challenges associated with implementing MKT tasks. Just as the mathematics tasks designed for children go through a series of transformations, MKT tasks similarly undergo change as they are taken up and used with teachers (see Figure 2).

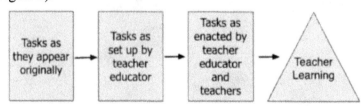

Figure 2: Mathematical Tasks Framework adapted to implementing MKT tasks with teachers

We have found that one of the main challenges of enacting MKT tasks is keeping the work focused on *mathematical* knowledge for *teaching*. Because MKT straddles both mathematics and teaching, it is easy for teachers' work on an MKT task to slide into an exclusive focus on either mathematics or teaching. For example, it is not unusual for a task intended to develop MKT to shift to mathematical work that is not necessarily connected to teaching. In the case of the $1\frac{3}{4} \div \frac{1}{2}$ task described earlier, the work could slip away from the development of MKT if it turned from a detailed discussion of the measurement and partitioning interpretations of division toward a more abstract discussion of Euclidean division in general. Alternatively, teachers' work could easily slide into purely pedagogical concerns without really engaging in the mathematics used in teaching if the discussion became focused on the typical trajectory for teaching fractions in schools or whether the standard algorithm for dividing fractions should be taught. While these mathematical and pedagogical shifts may result in rich explorations and interesting contributions to teachers' growth, losing a task's MKT-focus reduces the likelihood of developing the mathematical knowledge, reasoning, or skills that teachers use in instruction.

Another challenge of enacting MKT tasks is to make both the mathematics and the teaching salient. Teachers' compressed mathematical knowledge can make it difficult for them to see what there is to learn about a concept or to appreciate that connections between ideas are not obvious to those initially learning them. It can also be challenging to make visible and convincing to teachers that there is more to know than correct answers. For example, when working on the $1\frac{3}{4} \div \frac{1}{2}$ task, teachers might be skeptical of the need to represent the division of fractions using different interpretations of division. Teachers might also see some of the work on MKT tasks as needlessly complicating things that should be straightforward. For example, teachers may perceive attending to precision in language when writing division problems as obscuring what they consider a simple activity. Because of this, the opportunities that MKT tasks provide for "unpacking" the mathematics can be easily missed. Many aspects of the work of teaching may not be familiar or well known because so much of teaching practice is invisible (Lewis, 2007). Thus, connections between the work being done on MKT tasks and the work of teaching can easily be overlooked.

Keeping Tasks Focused on MKT During their Enactment

In our efforts to address these challenges, we have provisionally identified several guiding principles for keeping a task focused on MKT. One way to maintain this focus is to engage teachers in the work of attending to one another's thinking. This includes having teachers ask questions to clarify their colleagues' solutions, asking teachers to explain someone else's thinking, and encouraging teachers to figure out what might be confusing about the problem for someone else. This type of work is central to teaching and can involve a range of mathematical analysis, questioning, and reasoning.

A second way to focus teachers' work on MKT is to have them routinely explain their own thinking. This requires not only having them present and justify their solutions to others but also detailing what they find (or found) difficult and confusing. Although explaining one's thinking could seem central for work on any mathematics task, giving explanations is particularly important for teaching and developing MKT. Teachers often represent their own ideas and explicitly model their thought processes as they interact with students. Furthermore, talk is a main medium of classroom interactions, and thus teachers need opportunities to practice "talking mathematics."

Finally, we have found that it is important to be explicit about how a task is connected to the work of teaching. For example, the $1\frac{3}{4} \div \frac{1}{2}$ task can be connected to the work of analyzing students' errors. This diagnostic work is something teachers routinely do in practice. Connecting MKT tasks to the work of teaching can also involve asking teachers to engage in practices that are clearly part of what all mathematics teachers do—for example, providing opportunities to "talk mathematics" or record mathematical explanations and solutions on the whiteboard and other public writing spaces. Highlighting the mathematical demands of these mundane and seemingly generic teaching practices can also help teachers appreciate the specialized types of mathematical knowledge and skill used in their work.

Conclusion

Teaching is mathematically demanding work, requiring specialized knowledge, skills, and ways of reasoning not learned through typical mathematics courses or used in other professions. Our research group has been working to design tasks that provide teachers with opportunities to develop mathematical knowledge for teaching and to study the use and impact of these tasks. Through this work, we have begun to identify possible features of effective MKT tasks, as well as some of the challenges of enacting these tasks with teachers.

There is, however, still much to learn about designing and using tasks to teach mathematical knowledge for teaching. First, we need a better understanding of the features of MKT tasks. The list we propose here is admittedly preliminary, so it is likely that there are other key features. For example, particular aspects of MKT might develop more readily than others from repeated opportunities to practice using this knowledge and rehearsing specific skills. It could be that there may be certain concepts for which "provoking a stumble" is especially important.

Second, there is still much to learn about what is entailed in enacting MKT tasks. There is much to be learned from studies of the enactment of K-12 curriculum; the Mathematical Tasks Framework that was discussed earlier is a good example of this. And, just as teaching mathematics to children demands knowledge beyond that which is presented in the school curriculum, teaching teachers mathematical knowledge for teaching demands knowledge beyond MKT. As we better articulate the work of teaching MKT, we can also think about what knowledge and skills teacher educators need—i.e., *mathematical knowledge for teaching teachers* (MKTT) (Bass, 2007; Zopf, 2009).

Finally, it would be helpful to know more about how teachers learn MKT, and in particular, what can be learned from MKT tasks. As we become clearer about the design and enactment of MKT tasks, we will be able to study what teachers learn from their engagement in these tasks, as well as if and how this learning impacts their classroom practice.

References

Ball, D. L. (1988). *Knowledge and reasoning in mathematical pedagogy: Examining what prospective teachers bring to teacher education.* Unpublished doctoral dissertation, Michigan State University, East Lansing, MI.

Ball, D. L., & Bass, H. (2003). Toward a practice-based theory of mathematical knowledge for teaching. In B. Davis & E. Simmt (Eds.), *Proceedings of the 2002 annual meeting of the Canadian Mathematics Education Study Group.* Edmonton, AB: CMESG/ GCEDM.

Ball, D. L., Hill, H. C., & Bass, H. (2005). Knowing mathematics for teaching: Who knows mathematics well enough to teach third grade, and how can we decide? *American Educator, 29*(3), 14–17, 20–22, 43–46.

Ball, D. L., Sleep, L., Boerst, T., & Bass, H. (2009). Combining the development of practice with the practice of development in teacher education. *Elementary School Journal, 109*(5), 458-474.

Ball, D. L., Thames, M. H., & Phelps, G. (2008). Content knowledge for teaching: What makes it special? *Journal for Teacher Education, 59*(5), 389–407.

Bass, H. (2007, October). *Mathematical knowledge for mathematics education: Is it more than mathematical knowledge for teaching?* Presentation made at the Michigan Mathematics Education Leadership Conference, Ann Arbor, MI.

Begle, E. G. (1979). *Critical variables in mathematics education: Findings from a survey of the empirical literature.* Washington, DC: Mathematical Association of America and National Council of Teachers of Mathematics.

Lewis, J. M. (2007). *Teaching as invisible work.* Unpublished
    doctoral dissertation, University of Michigan, Ann Arbor.
Ma, L. (1999). *Knowing and teaching elementary mathematics:
    Teachers' understanding of fundamental mathematics in
    China and the United States.* Mahwah, New Jersey:
    Lawrence Erlbaum Associates.
Monk, D. H. (1994). Subject area preparation of secondary
    mathematics and science teachers and student achievement.
    Economics of Education Review, 13(2), 125–145.
National Mathematics Advisory Panel. (2008). *Foundations for
    success: The final report of the National Mathematics
    Advisory Panel.* Washington, DC: U.S. Department of
    Education.
Schifter, D., & Fosnot, C. (1993). *Reconstructing mathematics
    education: Stories of teachers meeting the challenge of
    reform.* New York: Teachers College.
Shulman, L. (1986). Those who understand: Knowledge growth
    in teaching. *Educational Researcher 15*(2), 4–14.
Stein, M., Smith, M., Henningsen, M., & Silver, E., (2000).
    *Implementing standards-based mathematics instruction: A
    casebook for professional development.* New York: Teachers
    College.
Zopf, D. A. (2009). *Knowing mathematics for teaching teachers:
    The mathematical demands of mathematics teacher
    education.* Unpublished doctoral dissertation, University of
    Michigan, Ann Arbor, MI.

Endnotes

[i] Earlier versions of this work were presented at the twelfth
    annual meeting of the Association of Mathematics Teachers
    Educators, January 2008, in Tulsa, OK and at the Research
    Presession of the annual meeting of the National Council of
    Teachers of Mathematics, April 2008, in Salt Lake City, UT.
    This submission is based upon work supported by the
    National Science Foundation under Grant No. 0455828. Any
    opinions, findings, and conclusions or recommendations
    expressed in this material are those of the author(s) and do not

necessarily reflect the views of the National Science
Foundation.

[ii] It is often mistakenly assumed that mathematical knowledge
for teaching (MKT) is meant to replace or provide an
alternative to Shulman's (1986) concept of pedagogical
content knowledge (PCK); however, we view this work on
MKT to be an elaboration of Shulman's work. The article by
Ball et al. (2008) in particular, provides a detailed discussion
of mathematical knowledge for teaching (MKT), including a
description of its relationship PCK.

[iii] One counterintuitive result that arises from the $1\frac{3}{4} \div \frac{1}{2}$
problem is that the quotient is larger than the dividend. When
working with division story problems, people are often most
familiar with stories involving whole numbers where the
divisors are greater than 1 (e.g., I have a dozen cookies
divided among four people; how many cookies does each
person get? Or, I have twelve yards of ribbon and need strips
that are two yards each; how many strips can I make?). In
these cases, the answer is smaller than the amount being
divided. Sometimes it surprises people to find that their
answer of $3\frac{1}{2}$ is greater than the dividend of $1\frac{3}{4}$.

Kara Suzuka is a researcher at the University of Michigan. Her
primary work focuses on developing practice-based educational
materials and approaches for use in pre-service courses as well
as in professional development contexts for teachers of
mathematics. Her research and materials development work
involve studies of classroom practice, utilize multimedia records
of classrooms, and center on efforts to engage teachers in
important practices of teaching.

Laurie Sleep is a researcher at the University of Michigan
School of Education. Her research interests build upon her
experience as an elementary teacher and include designing and

studying ways to help prospective teachers develop mathematical knowledge for teaching.

Deborah Loewenberg Ball is dean of the School of Education and William H. Payne Collegiate Professor at the University of Michigan. Her research focuses on mathematics instruction, and on interventions designed to improve its quality and effectiveness.

Hyman Bass is the Samuel Eilenberg Distinguished University Professor of mathematics and mathematics education at the University of Michigan. His mathematical research spans various branches of algebra. In mathematics education he collaborates with Deborah Ball and her research groups, working on mathematical knowledge for teaching (MKT), the development of measures of MKT, and the teaching of reasoning and proof.

Jennifer M. Lewis is a researcher at the University of Michigan. Before earning her doctorate, she taught elementary and middle school. Her research and teaching interests center on classroom practice as a site for teachers to learn mathematical knowledge for teaching.

Mark Hoover Thames is a researcher in the School of Education at the University of Michigan. His research focuses on practice-based and discipline-grounded approaches to studying content knowledge for mathematics teaching. His interests include research on teaching and teacher education, the development of mathematical literacy in under-served populations, and the mathematical education of teachers.

Superfine, A. C. and Wagreich, P.
AMTE Monograph 6
Scholarly Practices and Inquiry in the Preparation of Mathematics Teachers
© 2009, pp. 25–42

3

Developing Mathematics Knowledge for Teaching in a Content Course: A "Design Experiment" Involving Mathematics Educators and Mathematicians

Alison Castro Superfine
Philip Wagreich
University of Illinois at Chicago

The field of mathematics education currently lacks models for the design and implementation of content courses aimed at developing preservice teachers' mathematics knowledge for teaching. Our goal is to make visible the process employed by an interdisciplinary group of mathematics educators, a mathematician, and graduate students as we collaboratively designed and implemented such a content course. Using data taken from the course, we illustrate our design research process and the ways in which we negotiate conflicting ideas about the course design. By doing so, our aim is to encourage other mathematics educators and mathematicians involved with teaching content courses to similarly engage in iterative cycles of design in order to create more effective learning environments for preservice teachers.

Given the increasing emphasis placed on teachers' content knowledge in mathematics (Kilpatrick, Swafford, & Findell, 2001; RAND, 2003), content courses are becoming an integral part of the mathematics preservice teacher education equation.

With few exceptions (e.g., Stylianides & Stylianides, 2008), there is little evidence, however, that preservice teachers learn mathematics in ways needed for teaching from content courses. Moreover, the field of mathematics education lacks models for the design and implementation of such courses. This has resulted in many content courses being designed without sufficient attention to both the pedagogical and mathematical concerns entailed in developing preservice teachers' mathematical knowledge for teaching. Thus, this paper aims to make visible the process employed by a group of mathematics educators and mathematicians as they collaboratively designed and implemented such a content course. We argue that our experience has broader implications for the design of mathematics content courses for preservice elementary teachers.

In collaboration with colleagues at other institutions over the past several years, we have designed a required content course for elementary preservice teachers (PST) at the University of Illinois at Chicago (UIC) that is specifically structured around two main strands—learning mathematics and connecting that mathematics to the work of teaching. The course is also designed to use these strands as sites to learn mathematics in ways needed for mathematics teaching.[i] In designing the content course, we have engaged in a series of "design experiments" (Barab & Squire, 2004; Brown, 1992) that involve designing and implementing sequences of tasks in the course and then examining how these innovations influence PSTs' learning.[ii] The findings are then used to inform the design of subsequent iterations of various task sequences in the content course (Barab & Squire, 2004). What is unique about our design research process in the context of the course is its interdisciplinary nature. The content course represents an ongoing collaboration and negotiation in which members of the planning group associated with the content course engage in weekly discussions and debate about how to design and implement course tasks and activities aimed at developing PSTs' mathematics knowledge for teaching.[iii] The authors are members of the planning group, which includes mathematics educators, a mathematician, and graduate students.

Our purpose in this paper is to discuss two interrelated dimensions of our work. One dimension involves the implementation of a series of design experiments to create a learning environment that effectively supports PSTs' development of mathematical knowledge for teaching. The other dimension involves the ongoing process of collaboration and negotiation between mathematics educators and mathematicians, groups that sometimes have conflicting ideas about what PSTs should learn. We argue that, taken together, these two dimensions have greatly contributed to the overall effectiveness of the course design as well as our own ongoing professional development.

We will describe our collaborative effort at design-based research as it applies to the content course for PSTs beginning with a discussion of the evolution of the course. Then we describe our process of negotiation in designing and re-designing the curriculum, drawing on our teaching experiences and certain data sources from our research related to the course. Finally, we discuss what we have learned from our design efforts, both about the course and about our own collaborative process, and the implications for the next iteration of the content course.

The Course Context

The content course in question is the first of two required content courses PSTs take during their freshman or sophomore years, and precedes the mathematics methods course that PSTs typically take during their senior year. The content course is designed around developing PSTs' mathematical knowledge for teaching. Some researchers argue that preservice mathematics coursework focuses too narrowly on preservice teachers' development of common content knowledge (i.e., knowledge that bankers, retailers, or nurses, for example, have to know, such as computing percentages) and not enough on their development of specialized content knowledge (i.e., math knowledge specific to teaching, such as analyzing common student errors)—a kind of mathematical knowledge that more closely reflects the work entailed in teaching mathematics (Ball,

Thames, & Phelps, 2008; Hill & Ball, 2004; Mewborn, 2003).
The content of the course includes place value and number
operations, fractions, proportional reasoning, aspects of number
theory, and other elementary concepts. The course also provides
PSTs with opportunities to develop their abilities to engage in
explaining, representing, and understanding and reacting to
mathematical thinking that is different from their own. A typical
semester includes 29 class periods that are each 120 minutes in
duration. In any given semester, 15–30 PSTs enroll in each of
the two sections of the course.

Ongoing Development of Content Course Curriculum

Given the primary aims of the content course (i.e., to
develop PSTs' mathematical knowledge needed for teaching),
we have been developing and modifying tasks that reflect this
purpose and that also contribute to the development of an
inquiry-based learning environment within the course. Arguably,
developing PSTs' mathematical knowledge for teaching entails a
different conception of what mathematics is and how it can be
learned, a conception that may be unfamiliar given PSTs' prior
mathematical experiences. By asking PSTs to anticipate unusual
solution methods, for example, we are assuming that
mathematics tasks can be solved using a variety of different
methods and not simply using one "correct" method. Similarly,
by providing opportunities for PSTs to evaluate others'
conjectures, we are assuming that students will make conjectures
about mathematical relationships and will have to justify and
explain their thinking. Both of these mathematical practices are
likely to be unfamiliar to PSTs as they may not have engaged in
them during their previous mathematics coursework, including
their own K–12 schooling experiences. Thus, our course is
designed to create a learning environment and a set of tasks and
activities that require PSTs to engage in these types of teaching
tasks.

In our design experiments, we have focused much of our
efforts on designing tasks around the idea of posing and refining
definitions in order that they become more precise, and thus

mathematically valid. Such a task is inherent in the work of teaching and involves considerable mathematical work (Ball et al., 2008). For example, one sequence of tasks involves even and odd numbers wherein PSTs pose working definitions and use these definitions to prove statements about even and odd numbers such as the sum of two even numbers is also even. The nature and type of PSTs' proofs is often determined by their definition of even and odd number. For example, defining an even number as a whole number that ends in 0, 2, 4, 6, or 8 as opposed to defining it as a multiple of 2 can lead to the production of two very different proofs. After this task, we connect PSTs' work to teaching by having them evaluate the validity and usability of definitions of even and odd numbers taken from actual elementary textbooks. This is followed by a discussion and analysis of a classroom video in which children are discussing different definitions of even and odd. In this discussion, PSTs are asked to consider the mathematical issues that the teacher must manage in the lesson. Throughout these three tasks, PSTs are encouraged to continually refine and make more precise their definitions of even and odd numbers. Another sequence of tasks focuses on generating definitions for factor and multiple, part of which we use to illustrate our collaborative design research process in the next section.

With regard to the learning environment in the content course, our aim has been to create an environment that is collaborative in nature, where PSTs can inquire about mathematics concepts and ideas, engage in mathematical practices, and negotiate meanings with their peers (Yackel & Cobb, 1996). Within this learning environment, PSTs typically solve problems by first working individually on a problem and formulating questions and choosing an appropriate strategy, then working collaboratively on the problem with their peers, and finally formulating a solution and, when appropriate, a generalized formula. This process is typically followed by whole class discussion and, when necessary, negotiation of what counts as an acceptable solution.

Because the course is offered every semester and the curriculum has been continually re-designed to meet the needs of

PSTs, our content course is well positioned to serve as a venue for design-based research (Brown, 1992). That is, the course creates a useful context in which to design an intervention (i.e., a set of related tasks) and carefully study its implementation for the purposes of informing future iterations of its design.

Content Course as Design Research

Briefly, design experiments, or design research, were developed as a way to formatively test and refine educational designs based on previous research (Collins, 1992). As part of design-based research, researchers engage in iterative cycles of design and implementation, the goal of which is to produce new theories and practices that influence teaching and learning environments. In this sense, researchers move beyond the observation of naturalistic settings to designing and adjusting educational interventions for the purposes of testing and generating theories about learning in naturalistic settings (Barab & Squire, 2004).

Toward this end, we have been aiming to create a learning environment that not only supports PSTs' development of conceptual knowledge in ways that prepare them for their future work as teachers, but also that makes their thinking and learning visible to us as both instructors and designers. To accomplish this goal, formative and summative feedback on the course have been integrated into the design, creating a system that enables us to evaluate the extent to which we have been accomplishing our goals. In the next section, we provide an example of how we have been carrying out our design research process in the context of the course and creating a learning environment with opportunities for PSTs to learn mathematics for teaching.

Illustrating our Design Research Process

This example draws from our work on the content course during the Fall 2007 semester. We first focus on our overall design research process and then consider a specific task from the content course and the design considerations related to its implementation.

Documenting Course Activities

In order to examine our course design and its influence on PSTs' learning, we have been carefully documenting course activities across each iteration of the course. We have been videotaping each class session and randomly selecting focus PSTs for whom we audio tape class discussions in both small and whole group situations and collect any artifacts created by them during small group work. In addition, for each iteration of the course, one group member has been recording field notes of what transpires during class sessions, particularly focusing on PSTs' misconceptions that arise during class discussions, as well as recording any instructor reflections immediately following class sessions. The purpose of such extensive records has been to facilitate reflection on previous classes during weekly planning and debriefing meetings and also to document the overall course evolution. Such extensive records also serve as formative, and ultimately summative, feedback that has both enabled us to tailor course activities according to PSTs' emerging needs and informed our instructional decisions.

Planning and Debriefing Meetings

As part of our design research process, we have been meeting as a planning group on a weekly basis for about three hours each week.[iv] Generally, our meeting format is as follows. First, we debrief the previous week's classes in order to understand the overall "success" of different tasks and activities. We discuss why and how planned activities differed from enacted activities, with particular attention paid to the time used for each component of an activity. To determine PSTs' engagement in the course we examine interactions with peers and knowledge development. Class dynamics are assessed by considering how PSTs are participating in class discussions. We also take into account the ongoing development of the overall learning environment. Finally, in the latter part of our meetings, we plan for the following week, based on both the previous week's classes and our overall course goals. As we plan, we try to anticipate PSTs' misconceptions, develop activities that will elicit misconceptions, and come up with instructional strategies to address the misconceptions. Notes are taken, and these meetings are typically audio taped.

Negotiating Conflicting Ideas

An important part of our design process has involved negotiating conflicting ideas between group members about how to structure PSTs' learning opportunities in the course. We will use an example taken from our work on a specific task to illustrate how we have facilitated this negotiation process. For this paper, we draw only on transcripts of videotaped classroom sessions and notes and audio taped recordings from our planning and debriefing meetings. The specific task, the Locker Problem, requires students to show their understanding of the relationship between factors and multiples. The task is as follows:[v]

> There are 1000 lockers and 1000 students in Whitney Young High School. The lockers are numbered 1 to 1000. Student 1 runs through and opens every locker. Next, Student 2 changes the state of lockers 2, 4, 6, 8, and so on. Then, Student 3 changes the state of every locker numbered 3, 6, 9, and so on. Which locker doors are opened when every student finishes? If this pattern continues, what do you think Student 4 does?

As a planning group, we selected this task for a number of reasons. First, given that it was the beginning of the semester, we wanted PSTs to work on a task that entailed the kind of work and mathematical practices that were going to be emphasized throughout the course. Second, the task afforded a number of ways of recording data in order to find patterns leading to a conjecture. Finally, the task highlights the role of patterns in mathematics and creates an opportunity to discuss the conditions under which a pattern will continue indefinitely. In addition to involving the concepts of factor and multiple, we agreed that the task offered an opportunity for PSTs to formulate and test conjectures and to construct arguments to prove or disprove their conjectures—mathematical practices that were central to the course curriculum. However, we disagreed about whether or not to introduce a definition of factor and multiple *before* PSTs began working on the task. The mathematician and one graduate student questioned the efficiency of having PSTs work on this task without being given a definition at the outset. Instead, they suggested that we provide PSTs with the definitions of factor and multiple first and follow that up with applications of those concepts. In response, two graduate students and a mathematics educator argued that PSTs could begin working on the task and formulate hypotheses about which lockers would be open and closed without having a definition of factor and multiple and that students could develop and negotiate a working definition of these terms by working through the task. Furthermore, one mathematics educator argued that having these definitions arise in a natural way, in the context of solving a meaningful task, was a positive feature of the lesson design because it could provide an opportunity to discuss the necessity of precise language in mathematical definitions.

This discussion raised additional questions for members of our planning group: (1) how can we know whether this task is an efficient use of class time given the limited number of class sessions in a semester, (2) how do we know that the "traditional" way of initially introducing definitions of relevant concepts is an efficient use of time, and (3) what is gained if we introduce the task in a non-traditional way and allow students to negotiate their own definitions? With these two different task scenarios presented, we considered their benefits and limitations for PSTs' understanding of the relationship between factors and multiples. Overall, we agreed that it was necessary to consider the goals of the task in relation to PSTs' current level of understanding, which did not include formal class definitions of factor and multiple. While most students had some pre-existing ideas about the notions of factor and multiple, these were quite imprecise and often interchanged.

We ultimately decided not to introduce formal definitions of factor and multiple prior to working on the task as a way of

testing our ideas that by doing so 1) PSTs could negotiate their
own working definitions, and 2) the instructor could create an
opportunity to discuss the necessity of precision in mathematical
language. The following excerpt is taken from the Locker
Problem class that followed our planning meeting discussion.
PSTs, having worked collaboratively on the task for several
minutes, formulated conjectures about which lockers are open
and closed, as well as definitions for prime numbers and factors.

Instructor: So we have something about multiples.
                Anybody, other conjectures about
                multiples or...? Any other types of
                numbers?

Student 1:   Whether they're prime will only be
                open—will only be touched twice.

Instructor: Okay which ones?

Student 2:   Lockers that are prime.

Instructor: Others? What do—okay others? So that
                [concept is] new for we haven't actually
                talked much about [it] in the class right?
                Prime. So what about a working
                definition for the word prime? So prime
                number. What's a prime number? Um
                [Student 1] since you brought it up.

Student 1:   Okay, um, a number that is divisible by
                one and itself.

Instructor: Okay. What do you guys think? A prime
                number is any number, any, um,
                counting number that can be divisible by
                one and itself. Any number...I'm just
                gonna add an exactly, only. You said
                any number that's divisible by only one

and itself. Okay. So I just added only just to kind of reiterate that point. So lockers that are prime will only be touched twice. What about other conjectures? Other patterns you guys saw or things you were thinking about opening up those lockers? We will take you know one more. Yeah [Student 3].

Student 3:   The numbers of factors — the factors of the number say how many times it's gonna be touched.

Instructor: So that's another term, factors. What's a factor? Someone else besides [Student name] or [Student name]. What's a factor? What were you gonna say [Student name]? Go ahead.

The discussion continues for a few minutes during which Student 3 poses and then makes several attempts to revise her definition of factor. The instructor then revoices Student 3's definition:

Instructor: You said something about factors are the numbers that are multiplied to get that number. Okay so I think — so six for example. What are the factors of six?

Student 4:   One, two, three, and six.

Instructor: You guys agree? Disagree? Okay. So how can we — how can we say what factors are? Factors. Ideas?

Student 5:   The numbers that are multiples — or not multiples.

Student 6:   Yeah so okay, so factors is this weird
                term. Okay so the factors of six, they're
                like factors of a, when you're talking,
                okay so when we're talking about
                factors we aren't dealing with
                remainders at all.

Instructor: Right, so it's six — so one, two, three
                are factors of six. So there's something
                about factors being evenly divisible by
                or you know no remainders when
                they're multiplied to get the number.
                You see what I'm saying? Probably not.
                Okay. What did we say? Factors are...a
                number.

Student 7:   So it's easier to articulate the example
                like you do but then kind of come up
                with um exactly how to say it. Does
                anybody have an example?

Student 6:   Like 6. Factors are like the counting
                numbers that make up that number.

Student 3:   Factors of a number are the counting
                numbers by which the number evenly
                divisible by.

Instructor: By which the number is evenly divisible
                by. We'll refine this definition on
                Thursday...

As evidenced in the class excerpt, a definition of prime
number and factor did arise from the class discussion around the
Locker Problem without formal introduction by the instructor. In
fact, a definition of prime number also emerged from the
discussion, as the transcript illustrates. However, despite our
goal of using this task implementation as a way to have students

negotiate working definitions and discuss the necessity of precise language when formulating definitions, these opportunities did not arise in the remainder of this class session, nor did they emerge in the subsequent class session when PSTs finished their work on the Locker Problem.

This episode, albeit short, raised a number of questions and considerations in the planning and debriefing meeting that followed this class session. First, the instructor in the Locker Problem vignette made the decision to sharpen the prime number definition posed by Student 1 by adding "exactly one" to the definition posed, "a number divisible by one and itself." When asked about this interjection, the instructor said she decided to clarify and pose a valid definition for prime number in order to move on to factor and multiple, which was her primary goal for the class session. While one group member agreed that this was a good decision in terms of curriculum coverage and time constraints, several of the other group members disagreed, stating that this instance was a missed opportunity for PSTs to engage in an important mathematical task of teaching. PSTs did not have the experience of negotiating a precise definition of prime number, and in particular, gaining an understanding of why a definition without the word "only" would not suffice to define "prime."

Nevertheless, the instructor similarly decided not to negotiate working definitions of factor and multiple because she was convinced that PSTs understood these two concepts once the definitions were posed. This decision caused us to consider what we take as evidence of understanding, which led us to a hypothesis that PSTs may exhibit different levels of understanding of these concepts. For example, PSTs could exhibit an instrumental understanding (Skemp, 1976) of factor that would permit them to answer procedural questions such as, is 7 a factor of 56? On the other hand, PSTs could exhibit a more relational understanding of factor that would support their ability to explain conceptual questions such as, if 7 is a factor of n then n divided by 7 is also a factor of n. Overall, the questions raised from our planning discussion of the Locker Problem have framed subsequent discussions about the course content and the ways that such content should be taught in subsequent classes as well as future iterations of the course.

Our planning discussion about how to structure and implement the task was extremely valuable to both the course instructor and the other members in the planning group. Fleshing out alternative strategies and their underlying rationales in a collaborative forum that included a mathematician, mathematics educators, and several graduate students significantly helped us (re) design what seemed to be an effective lesson for PSTs. Moreover, this sort of collaboration provided an important opportunity for the professional development of every member of our group; we were encouraged to analytically consider alternative disciplinary perspectives in light of the goals of the course.

Implications for Our Next Iterative Cycle

Given the outcome of PSTs' work on the Locker Problem and other evidence indicating that PSTs could articulate the relationship between factors and multiples upon completing the task, we have decided to introduce the Locker Problem in a similar fashion in future course iterations, allowing the definitions of factor and multiple to arise naturally from the task discussion. We have been discussing this issue of initially introducing definitions in the context of other tasks. For example, another task involves proving addition and multiplication statements about even and odd numbers. We hypothesize that agreeing on definitions of even and odd numbers prior to working on the task would profoundly influence the proofs that PSTs would generate. For example, using an algebraic definition (e.g., an even number is a number of the form $2n$) may lead to an algebraic proof, while a definition based on groups of 2 or partitioning into two equal groups can lead to visual, verbal, or symbolic proofs of great variety. Still, another task involves explaining why certain divisibility rules work. Would posing a definition of what it means to be divisible influence PSTs' explanations of the divisibility rules? For example, what are the implications if PSTs use the following definition of divisible: x is divisible by 9 if $x = 9n$, where n is an integer? Such considerations are integral to our design work around the content course.

Conclusion

We began this article by arguing that PSTs often do not develop a strong knowledge of mathematics, particularly of the sort needed to teach elementary mathematics. In our introductory content course for PSTs, we have been engaged in iterative cycles of design research with the aim of developing a course curriculum and learning environment that supports PSTs' understanding of complex mathematics in ways needed for teaching. Using specially designed tasks, content sequencing, and assessments, we have been continually re-designing the content course based on knowledge gained from past iterative cycles of development. Moreover, we have used this cycle as a foundation for considering alternative disciplinary perspectives in a way that is informed by evidence about a common object of interest and inquiry. By making our design process more visible, we hope to encourage other mathematics educators and mathematicians involved with teaching content courses to similarly engage in continual cycles of design and re-design, always reflecting on what was learned in one semester and applying this knowledge to improving subsequent iterations of the course. Through this design-based research process, important ideas and insights can be applied to the design of effective learning environments in ways that support PSTs' learning of mathematics needed for teaching.

References

Ball, D., Thames, M., & Phelps, G. (2008). Content knowledge for teaching: What makes it so special? *Journal of Teacher Education, 59*, 389–407.

Barab, S., & Squire, K. (2004). Design-based research: Putting a stake in the ground. *Journal of the Learning Sciences, 13*, 1–14.

Brown, A. (1992). Design experiments: Theoretical and methodological challenges in creating complex interventions in classroom settings. *Journal of the Learning Sciences, 2*, 141–178.

Collins, A. (1992). Toward a design science of education. In E. Scanlon & T. O'Shea (Eds.), *New directions in educational technology* (pp. 15–22). New York: Springer-Verlag.

Hill, H., & Ball, D. (2004). Learning mathematics for teaching: Results from California's mathematics professional development institutes. *Journal for Research in Mathematics Education, 35*, 330–351.

Kilpatrick, J., Swafford, J., & Findell, B. (Eds.). (2001). *Adding it up: Helping children learn mathematics*. Washington, DC: National Academy Press.

Lappan, G., Fey, J., Fitzgerald, W., Friel, S., & Phillips, E. (1998). *Prime time*. Menlo Park, CA: Dale Seymour Publications.

Mewborn, D. (2003). *Teaching, teachers' knowledge, and their professional development*. In J. Kilpatrick, W. G. Martin, & D. Schifter (Eds.), *A research companion to Principles and Standards for School Mathematics* (pp. 45–53). Reston, VA: National Council of Teachers of Mathematics.

RAND Mathematics Study Panel, D. Ball, Chair (2003). *Mathematical proficiency for all students: Toward a strategic research and development program in mathematics education*. Arlington, VA: RAND.

Skemp, R. (1976). Relational understanding and instrumental understanding. *Mathematics Teaching, 77*, 20–26.

Stylianides, A., & Stylianides, G. (2008). Studying the implementation of tasks in classroom settings: High-level mathematics tasks embedded in real-life contexts. *Teaching and Teacher Education, 24*, 859–875.

Yackel, E., & Cobb, P. (1996). Sociomathematical norms, argumentation, and autonomy in mathematics. *Journal for Research in Mathematics Education, 27*, 458–477.

Endnotes

[i] The first author was formerly a member of a group of mathematics educators, mathematicians, and graduate students who developed methods and content courses for elementary preservice teachers at the University of Michigan. Some of the ideas used in the UIC content course grew from this collaboration.

[ii] We use the phrases "design experiments," "design research," and "design-based research" interchangeably throughout the text.

[iii] Different members of the planning group serve as course instructors during different semesters.

[iv] Members of the planning group include mathematics educators, a mathematician, and mathematics education and learning sciences graduate students.

[v] We used a version of the Locker Problem based on materials developed by colleagues at the University of Michigan–Dearborn. Another version of the Locker Problem can be found in Lappan et al. (1998).

Alison Castro Superfine is an assistant professor of Mathematics Education and Learning Sciences in the Department of Mathematics, Statistics, and Computer Science at the University of Illinois at Chicago. Her research interests include elementary preservice teacher education, teachers' mathematics knowledge, and teacher-curriculum interactions.

Philip Wagreich is a professor of Mathematics in the Department of Mathematics, Statistics, and Computer Science at the University of Illinois at Chicago. His research interests include algebraic geometry and mathematics teacher education.

Author Note: We would like to acknowledge and thank Kelly Rivette, Sarah Oppland, Kathleen Pitvorec, and Wenjuan Li for their many contributions and insights, which have continually shaped the content course and have helped to build a foundation for our design work.

Dixon, J. K., Andreasen, J. B. and Stephan, M.
AMTE Monograph 6
Scholarly Practices and Inquiry in the Preparation of Mathematics Teachers
© 2009, pp. 43–66

4

Establishing Social and Sociomathematical Norms in an Undergraduate Mathematics Content Course for Prospective Teachers: The Role of the Instructor

Juli K. Dixon
Janet B. Andreasen
University of Central Florida

Michelle Stephan
Lawton Chiles Middle School and University of Central Florida

The establishment of social and sociomathematical norms in elementary classrooms has been well established. Paying explicit attention to the development of these norms in an undergraduate mathematics content course for elementary school teachers led us to explicate the role of the instructor in establishing these norms. While conducting a classroom teaching experiment, we identified that the role of the instructor can be organized into three phases, namely (a) planning for negotiation, (b) negotiating new norms, and (c) sustaining norms.

Reform movements in mathematics education at all levels provide substantial support for teaching mathematics with understanding (National Council of Teachers of Mathematics (NCTM), 2000). The importance of establishing normative ways of interacting with the K–12 classroom community, which foster these types of mathematical understandings, has been well documented (Cobb, Wood, Yackel, & McNeal, 1992; Stephan & Whitenack, 2003; Yackel, 2001; Yackel & Cobb, 1996). In undergraduate mathematics classrooms, Yackel, Rasmussen, and King (2000) documented social and sociomathematical norms in a differential equations course. They explicated the importance of establishing specific types of social and sociomathematical norms that promote student learning. They were able to extend to the college level "the body of work at the elementary school level that shows how norms characteristic of inquiry instruction contribute to the conditions which make meaningful mathematical learning possible" (Yackel et al., 2000, p. 276). The study presented here sought to contribute to the body of knowledge at the college level by examining the process by which social and sociomathematical norms might be established within the context of an undergraduate elementary mathematics content course for prospective teachers. This environment is different than the environment examined by Yackel et al. (2000) due to the fact that as prospective teachers explore their own understanding of elementary school mathematics, they simultaneously develop strategies for providing similar learning situations for their future students. While the instructor and prospective teachers negotiate norms for operating in a classroom environment, the instructor is modeling ways to establish norms that the students might implement in future settings when they assume the role of instructor. Therefore, this paper could be helpful as mathematics teacher educators plan for and set up their social environments and sociomathematical norms in mathematics content courses for prospective teachers.

Social and Sociomathematical Norms

Social norms refer to accepted ways of participating in the classroom community. While social norms occur in every classroom, our focus in this chapter is on the participatory structure that supports an inquiry learning environment. Social norms shown to be important in inquiry classrooms typically include that students develop meaningful solutions to problems, explain and justify solutions and solution processes, attempt to make sense of other student's solutions, and ask questions or raise challenges when there are misunderstandings or disagreements (Yackel and Cobb, 1996). The classroom sociomathematical norms involve criteria for what counts as a different, unique, efficient, or sophisticated mathematical solution as well as what counts as an acceptable mathematical explanation and justification (Yackel, 2001). Social and sociomathematical norms are established by the teacher and the students cooperatively. Although the teacher is an authority figure in the classroom, the teacher can only initiate and guide the process of establishing the social norms. The teacher cannot demand that specific norms be established as the students in the classroom must be involved in negotiating norms (Cobb, 2000). Research on social and sociomathematical norms has been conducted primarily with elementary children, with a notable exception being Yackel et al. (2000). Not only are we interested in how teachers establish norms with adult students but also with pre-service teachers who must then try to emulate this with their own students some day.

Methodology

This study used qualitative research methods to document a classroom teaching experiment conducted in a semester-long undergraduate mathematics education course for prospective elementary school teachers. The design of a classroom teaching experiment allows the researchers to experience firsthand the ways in which students learn and reason about mathematics. "Students' mathematics is indicated by what they say and do as they engage in mathematical activity, and a basic goal of the researchers in a teaching experiment is to construct models of students' mathematics" (Steffe & Thompson, 2000, p. 269). The teaching experiment is designed as a sequence of teaching episodes which include a teacher, one or more students, at least one witness of the teaching episode, and a method for recording the teaching episode. The sequence of teaching episodes can span anywhere from several class sessions to entire courses, in this case a semester-long undergraduate course in elementary mathematics. The goal is to construct models of students' mathematics that can be useful in teaching other students and in learning how students learn and understand mathematics. In the classroom teaching experiment reported here, the main goal was to understand how prospective teachers develop mathematical understandings of number concepts and operations with an underlying goal of examining the establishment of social and sociomathematical norms in this context.

Participants and Instructional Setting

Participants in this study were undergraduate students at a large university located in the southeastern United States and were predominately prospective elementary teachers in their sophomore or junior year. All 16 students enrolled in the course agreed to participate in the study. Students were seated at tables of four to five students each and were given time on most tasks to work individually and with their small groups before a whole class discussion was begun. The course met three times a week for approximately three hours per session during a 6-week summer session. While the focus of the course was on number, geometry, and measurement, the course began with a short problem-solving unit. A context of a candy shop where packaging of candies is situated in base eight was used as a basis for the instructional sequence for whole numbers while instruction in fractions was situated in base ten (Andreasen, 2006; Wheeldon, 2008). While geometry and measurement were part of the course, the teaching experiment did not include these instructional sequences which made up the last week of the course.

Data Collection and Analysis

Each class session was videotaped throughout the teaching experiment. The research team recorded field notes to further supplement the videotapes. At the conclusion of each class, each member of the research team kept a journal to record their perspectives of the sequence implementation, classroom dynamics, and individual student development (Steffe & Thompson, 2000). The research team met after each class session to discuss the day's results and plans for the next session (Cobb, 2000; Simon, 2000). These meetings were audio taped.

We examined classroom interactions to document the establishment of social and sociomathematical norms using the methodology proposed by Cobb and Whitenack (1996), in which videos are first analyzed chronologically to identify instances that appear to assist in establishing classroom social and sociomathematical norms. These chronological instances were then correlated to different norms. Finally, these instances were synthesized to determine social and sociomathematical norms that were taken-as-shared by the classroom community. We also made conjectures about the establishment of norms based on field notes and observations and compared these to the conjectures made in analyzing the video transcripts. Times when it appeared that norms were being established were marked and then later reviewed to look for instances of consistency between class sessions and/or inconsistencies that might indicate that a classroom norm was not taken-as-shared. For example, a student might violate what we thought was a norm of the classroom, but because other students did not object, we had to reconsider whether the norm was taken-as-shared.

We determined that two social and two sociomathematical norms were established:

Social norms

1) explaining and justifying one's solution and solution processes
2) making sense of other students' solutions by asking questions of classmates or the instructor

Sociomathematical norms

1. what constituted a different mathematical solution
2. what made a good explanation.

These norms were negotiated between the students and the instructor and constituted what was acceptable participation socially and mathematically (Cobb, 2000; Stephan & Whitenack, 2003). The established norms supported the ways that students communicated with each other and with the instructor as well as contributed to the mathematical understandings of the community. In an effort to document the process used in establishing these norms, themes related to this process were then examined in both the video tapes of class sessions and the audio tapes of research team meetings using Glaser and Strauss' (1967) constant comparative method. Specific roles of the instructor, the research team, and the students were of particular interest. Representative excerpts were then selected to illustrate the methodology for establishing norms.

Results

The video and audio transcripts were analyzed with the goal of unpacking the process by which social and sociomathematical norms were established. This process was organized according to three phases of the experiment, namely (a) planning for negotiation, (b) negotiating new norms, and (c) sustaining norms.

Planning for Negotiation

Most of the data related to the theme of planning for negotiation were identified through analysis of audio transcripts from the research team meetings and showed that a large part of the process of initiating the negotiation of norms included careful planning prior to teaching and then adjusting that plan in action to capitalize on opportunities that lent themselves to the norming process. This finding is not surprising; however, unpacking what it means to plan for the negotiation of norms is helpful in establishing classroom environments that promote inquiry discourse. Questions that needed to be addressed were as basic as, "What norms are important when addressing mathematical knowledge for teaching?" "What does it mean to 'negotiate norms' with adult learners?" "How does one initiate the negotiation process?" In this study, planning for the quality of discourse that enables norm building involved three components: (a) placing norms in the foreground, (b) developing methods for eliciting student responses, and (c) providing opportunities for negotiation.

Placing norms in the foreground. Our first research team meetings focused on discussing the instructor's strategies for establishing norms. Establishing norms at the start of the semester was important to the team because we felt that if students did not have a safe environment in which explaining and justifying one's thinking was an obligation, then opportunities for learning mathematics with understanding would be minimal. To this end, we started instruction with problem solving tasks that did not require sophisticated mathematical understanding of particular content. While students were developing solutions to these problems, the teacher used their reasoning as the springboard for establishing norms for communication and participation. In the following sections we make explicit the role the instructor played in creating an interactive environment among prospective teachers.

Developing methods for eliciting student responses. When trying to create a participatory environment with pre-service teachers, one of the more difficult social norms to establish was the obligation to make sense of each other's explanations. In the research team meeting following the second class session, a discussion ensued as to how the instructor might support student discourse that would lead to the establishment of this norm.

> *Observer:* There was some student discourse tonight, but it was all, let me present, sit down, present, sit down. There was no time or opportunity to analyze each other's thinking, accept each other's thinking.... There was student discourse tonight, but what we were suggesting, we might change the nature of the discourse now, in light of this discussion, to not just presenting our answers but having an opportunity to analyze each other's.

Instructor: I didn't know how to get to that next level. And I was trying to, and I couldn't come up with a question. So, could you repeat the question to get that to happen? An example of a question to get that to happen?

Observer: I don't have to give you a question. I'll say, Jessica, re-explain what Alex just did. You understand what Alex just did? Just re-explain it for us, because I think I might have missed it. And she re-explained it, and you can tell whether she understands what he did or not. And then, you might ask something like, well Zachary, what do you think about that way? Was that the same way as yours, or different way than yours? Do you think it was easier or harder than your way? Analyze it.

Instructor: And I think that some of that can come up naturally. And I think what was happening tonight was students who say, "Oh, I get it." And that was it.

As this excerpt shows, our initial attempts to support productive interactions in class were not as successful as we hoped. Generally, students' discourse tended to stay at the level of sharing their strategies with no attempt by others to analyze what they had just heard. Getting students to make judgments about solution strategies and feel obliged to understand others is difficult at all ages. To change the way students reacted after a sharing session, the instructor made a concerted effort to ask other students to (1) re-explain what they just heard, (2) make judgments about the strategy, and (3) compare their own strategy with the one they just heard. Then, the instructor began to call on specific students to reflect on other students' methods with the intention of providing opportunities for students to recognize and fulfill their obligations to make sense of others' thinking.

Providing opportunities for negotiation. While helping students learn how to reflect on and understand their peers' solutions, another norm that must be negotiated is what counts as an acceptable explanation. If students are not giving productive explanations, their peers will not see the need to understand each other. In our teaching experiment, there were questions in both the research team meetings and by students in the class as to what counted as acceptable. Our goal was that the instructor would assist the students in defining these criteria for themselves. As students prepared to complete their first homework assignment in which they were asked to provide explanations and justifications, concern arose on the part of the students as to the expectations for written explanations. The instructor desired that the criteria be negotiated with the students and not be imposed by the instructor; therefore, this needed to be explicitly addressed in the class and thus required planning. Part of that planning involved the research team members discussing what criteria should be present, as in the dialogue that follows.

Observer 1: I was marking where I saw evidence of good explanations. I was trying to make a note that seemed like a good explanation.

Observer 2: What did you use as your criteria?

Observer 1: That it was complete, and gave enough detail that other students in the class could follow. And could then later explain what they did for themselves. I guess enough detail was a big part of it for me, that they – they explained the mathematics in a way that showed that they understood what was going on.

Observer 2: I say each step that I did, and I say why I did each of those steps and I try to relate it to … for me there is a difference between a written explanation and the criteria for those, versus a verbal one. What gets constituted as criteria for

verbal explanations may change during the
course — does change during the course of the
semester.... But for a written one. Think about
mathematical proofs in journals, and they have
certain forms and there are criteria mathematicians
just in our discipline have developed for what counts
as an acceptable proof.

This excerpt shows the research team struggling to articulate
what counted as a good explanation. As a result of the discussion
that took place, the research team established criteria for what
counted as both acceptable written and oral explanations, which
included providing step-by-step procedures along with the
mathematical justifications for why those procedures were
appropriate. Once criteria for what counted as an acceptable
explanation were established within the research team,
experiences for the students were planned so that the teacher and
students could begin to co-establish norms for participation in
this classroom. Although the focus of this excerpt was on written
explanations, the same discussions ensued regarding verbal
explanations.

Negotiating New Norms
     It is important to think ahead regarding the negotiation
process. Teaching moves for establishing norms should be
intentional. Regardless of the level of planning, however, the
classroom environment will be influenced by the prospective
teachers who are enrolled in a given semester. While the
research team *planned* for a certain type of negotiation,
transcripts of class sessions were analyzed to determine the
nature of the social and sociomathematical norms that were
actually established. In the process of this analysis, two themes
emerged: (1) initiating the shift in responsibility for learning and
(2) explicating norms.
     Initiating the shift in responsibility for learning. As norms
were negotiated, two shifts occurred related to student
responsibility: a shift in the obligation for explanations and a

shift in the centrality of authority. The instructor's role was pivotal in facilitating these shifts.

Shift in obligation for explanations. As the norms were negotiated, there was a shift from the instructor asking for explanations to the students providing explanations on their own. Initially, the instructor provided the impetus to supply explanations by explicitly requesting them. In this episode, which occurred during the first class session, students were working to find configurations of 16 coins (quarters, nickels, and dimes) worth $1.85. The instructor prompted students for explanations of their solutions.

Instructor: Did anyone get another solution?

Sarah: Four quarters, five dimes, and seven nickels.

Instructor: How many got that one? How did you get it Sarah?

Sarah: Well, I started with the quarters first. It was just basically all trial and error. Well, I knew that if I had four quarters, I would have to have an odd number of nickels. And so I just played around with the dimes. I went from like five nickels to seven nickels.

Sarah initially provided her answer but needed prompting from the instructor to supply an explanation. Because the planning team was concerned about emphasizing the expectation that students would supply explanations without prompting, we decided that the instructor would explicitly state the importance of explaining and justifying one's answer in class. As students were solving a problem involving using a deck of cards to find the least common multiple of 2, 3, 4, 5, and 6, a student, April, provided an explanation for her solution. The instructor took the opportunity to point out that providing explanations and justifications for solution processes was an expectation for all the students in the course. She also explicated the distinction between an explanation and a justification. This distinction was negotiated well into the semester.

Instructor: That *why* is what we are going to be focusing on in the content that we approach during this semester. This is actually a very interesting problem and one that we can spend an awful lot of time exploring the why we get down to the prime factors —that if we multiply the prime factors we end up with the least common multiple for the numbers… That is what we are going to be doing with the content of this course. Exploring that *why*. What I want to be able to do and what I want you to have as a responsibility to do as we are dealing with the content in the course is to always be able to say why. So when we talked about how we got this answer, we were focusing on the procedures. "This is what I did – I broke them down into prime factors. I made sure that each prime factor for the problems was represented in my final solution and when I knew that all the prime factors were represented in my solution, I could multiply them together to get my least common multiple for these numbers." That is the procedure. Why it works is what we are going to focus on – the justification for why I can use this procedure and get the right answer.

This discussion paved the way for further negotiations of the need to provide explanations and justifications as the course continued. Additionally, the instructor explicated the intent for students to take the responsibility for providing explanations and justifications, thus beginning the shift in responsibility from the instructor to the students. As the semester continued, students provided explanations and justifications without being prompted and there was less need for the instructor to explicitly state expectations for student behaviors in the classroom community. In the episode below, students were solving a problem in which a machine was placing 10_8 sticks of gum in each pack. They wanted to find out how many sticks of gum were in five packs.

Carrie provided her explanation and justification, which was based on repeated addition, along with her answer without prompting, recognizing her responsibility to provide the explanation automatically.

Carrie: I wrote 10_8 five times because there was five packs and each pack had 10_8 pieces.

Instructor: Okay.

Carrie: I just added them. I did 10_8 plus 10_8 is 20_8 and 10_8 plus 10_8 is 20_8 and I knew that equaled 40_8 plus 10_8 is 50_8.

This transcript provides evidence that students began to provide explanations on their own. The prompting of the instructor and the results of planning with respect to placing norms in the foreground were important aspects of the negotiation phase.

Shift in centrality of authority. As norms were established, there was a shift from the instructor as authority to the students as a more autonomous community of learners. At the beginning of the course, it was entirely the instructor's responsibility to ensure that students understood solutions and to ask questions when there was a need for clarification. As detailed above, initially the instructor requested explanations. In an effort to shift the responsibility for questioning from the instructor to the students, the instructor occasionally attempted to take the role of a student in order to model what was expected.

In this episode, David provided a solution to how to configure 246_8 candies into boxes of 100_8, rolls of 10_8, and individual pieces. The instructor prompted students to make sense of each other's solutions and, in the process, continued to emphasize the need for students to take the responsibility to ask questions. The instructor continued to take the role of the student in prompting what was expected participation in the classroom.

David: One box, 2 rolls, 116_8 pieces, [I mean] 126_8 pieces.

Instructor: Laura, how did David get that?

Laura: I couldn't tell you.

Instructor: Then you have some questions to ask David. Ask a question.

Laura: How did you get that?

David: Take the box and turn that into a 100_8 pieces. Take 2 rolls would be 26_8 pieces. So you have 1 box, 2 rolls, and 126_8 pieces.

Instructor: What did he do, Laura?

Laura: I see what he did.

Instructor: Explain it to me. I got a little lost.

It was expected that the students would ask questions if there was confusion. Because Laura could not explain David's answer, the instructor prompted her to ask questions directly to David. This episode demonstrates a technique the instructor used to encourage students to start asking questions of each other and, thus, to consider each other as authorities rather than the instructor. The authority in the classroom shifted from the instructor to the students with respect to both the responsibility to question and the responsibility to answer. Eventually, there were segments of dialogue that were entirely between students, such as in the following transcript. Students were making sense of a child's solution to the base ten problem, $47 - 29$, in which the child attempted to compensate but did so incorrectly. The child changed 47 to 50 and 29 to 30 and then found $50 - 30$ to be 20. Because he added 3 to 47 and 1 to 29, he added $3 + 1 = 4$ and said the solution was $20 - 4 = 16$.

Kelly:	If you do it later on... he added the 3 and the 1 he should have subtracted 3 and 1....
David:	I'm not getting it, because if you add the extra 2 and you get 52 minus 32 it would still be 20 and you are saying you do 20 minus 6?
Kathy:	No, we aren't adding 2.
David:	Let's say you did and you are using your formula, how would that go?
Kathy:	We aren't adding 2 to the 50 and we're not adding 2 to the 30.
Amy:	We are saying the difference between the 3 and the 1 is 2,
Kathy:	You are still subtracting.
Amy:	The difference between the 3 and the 1 is 2 which means that is what you are going to end up with the 20 minus something and you are going to subtract two because whatever you do to one number you should do to the other, but he did two different numbers.

The dialogue was completely between students in this episode, suggesting that the authority in the classroom had shifted to the students as a community of learners. The instructor's role in this shift was influenced by the planning for negotiation in placing norms in the foreground and defining strategies for eliciting student responses. The shifts in responsibility and authority were critical for the establishment of norms in the classroom community.

Explicating Norms

As the course progressed, the need to renegotiate the meaning of norms with the students became evident. Because the social norm of making sense of other students' solutions and the sociomathematical norm of what counts as an acceptable explanation were the most difficult to establish, it was not surprising that they needed renegotiation.

Making sense of other students' solutions. Initially, students seemed content to make sense of their own solutions, but did not feel obliged to make sense of other students' solutions to the same problems. To combat this problem the research team suggested that the instructor ask students to explain how other students solved problems. The following episode illustrates how students continued to think egocentrically when explaining solutions. Students were solving a problem involving reconfiguring 3_8 boxes, 6_8 rolls, and 4_8 pieces.

April:	I unpacked one of the boxes and got 10_8 rolls. I unpacked one of the rolls and got 10_8 candies out of that.
Instructor:	So what did you end up with?
April:	2 boxes, 15_8 rolls, 14_8 pieces.
Instructor:	Darren, how did she get 15_8 rolls in this?
Darren:	Each box contains 100_8 pieces, so then that would be 200_8 pieces plus the 15_8 rolls which each roll has 10_8 pieces when you break it down to each piece.
Instructor:	So it sounds to me like you took the box and broke it down into pieces. How many pieces?
Darren:	300_8
Instructor:	What I am asking you is how April got 15_8 rolls.

Darren: Well, she broke down one of the boxes into 10_8 rolls and then added it to the five existing rolls to equal 15_8 rolls.

Instructor: But there are 6 rolls.

Darren: She broke down one of the rolls into pieces.

Darren's response indicated he was making sense of his own solution rather than how April might have obtained her solution. When pressed, he was able to make sense of her solution for himself. The importance of this norm is two-fold in the context of prospective teachers. First, if prospective teachers only make sense of their own solutions, they may never be challenged to create more sophisticated ideas. Second, prospective teachers will one day need to be able to make sense of children's solutions in their own classrooms.

Acceptable explanation. During planning we used students' experiences writing explanations in order to establish criteria for what counts as an acceptable explanation. The team decided to promote an explicit discussion about what counts as an acceptable explanation by asking the students to examine a middle school problem in which three different explanations were provided. In this example, middle school students were asked to find out how many apples a family of 5 would eat in a year if they each ate 5 apples per week. In the first explanation, the student gave only the answer with a statement that he multiplied. In the second explanation, the computation was included along with the final answer. In the third explanation the student explained how he did the computation and why he chose the numbers he did and what the results of the calculations stood for in the story problem. The students in the class provided justification for why the third explanation was the most complete and most acceptable. They were able to describe that each subsequent argument gave more detail than simply an answer, and the third explanation provided the reasons for the calculations. The students were then asked to work in their small groups to create acceptable explanations for problems in base eight, one of which involved a picture of 4 rolls and 4 pieces of candy. Students were asked to determine how they might unpack candies to sell 27_8 pieces of candy and then to determine how much candy was left over. Kathy provided a written explanation, and students assisted in determining what was missing.

Kathy: You begin with four rolls, which equals 40_8 pieces, plus the four individual pieces. In order to sell 27_8 pieces, you must unroll three rolls, which now gives you 30_8 individual pieces. To determine how many candies will be left, we subtract 27_8 from 44_8 which is the total candies and we did the math....We approached it from the point of you had 27_8 pieces so we ... looked at that part first and said, well, if you have 4 rolls and 4 pieces, then you are going to have to break down at least three rolls to equal 30_8 pieces to get to 27_8.

Instructor: Questions? How did you know that 4 rolls gave you 40_8 pieces?

Amy: Because you know there is 10_8 pieces per roll.

Kathy: But we didn't say that.

Amy: So we should explain how many pieces per roll.

Kathy: We just said that 4 rolls which equaled 40_8 pieces.

With this task, the instructor provided opportunities for students to negotiate the criteria for the sociomathematical norm of what counts as an acceptable explanation. Their criteria involved reporting 1) the answer, 2) the calculations that led to the answer, and 3) why those calculations were used. The planning phase of establishing norms was critical in this negotiation.

Without planning for discourse, the negotiation of social and sociomathematical norms may not have been as successful.

Sustaining Norms

At some points during the semester, the focus shifted from establishing norms to maintaining them. This shift often occurred following an examination where the instructor was once again seen as the sole authority figure. Thus, the instructor must be intentional about reestablishing negotiated norms. The sequence of lessons on rational numbers began midway through the semester and followed the examination on whole numbers. The instructor began the first lesson on rational numbers with the following:

> We're starting fractions tonight and just as you thought very hard in eight world [referring to base eight problems in the previous sequence] ... fraction world is going to involve some thinking. So, once again, when you make statements be prepared to explain them...Hopefully you'll continue making statements and being involved in the discussion as you were in the other unit in base eight and base ten because you had some wonderful discussions going on.

Even with this reminder, there were several instances during the first half of the class period that the instructor needed to prompt students to provide explanations and justifications, using such prompts as "Tell me why," "How did you get it?" and "Why did you do that?" In the second half of that class session, however, explanations were often given without prompts. Once social and sociomathematical norms were established in this classroom community, the instructor's efforts shifted from negotiating to sustaining norms, a responsibility that seemed to be shared by the students. Classroom transcripts provided evidence that students actualized their responsibility with the instructor in the background providing prompts.

In addition to the shift in roles from the instructor to the students in sustaining social norms in the classroom, established sociomathematical norms continued to be supported as the semester progressed. Just as with social norms, it became evident that students took responsibility for maintaining norms of what counted as an acceptable explanation. In the following episode, Lilly explained how she solved the following problem: "Betty had 2 1/2 yards of ribbon. She gave 2/3 of a yard of her ribbon to Wilma. How much ribbon did Betty have left?" David asked a question to help make sense of Lilly's solution, providing support for making sense of one another's solutions. Matt enforced the criteria of what counts as an acceptable explanation by asking her to include why her procedures were acceptable.

Lilly: So instead of subtracting, I added to get my answer. By adding the remainder of what was left. And then I added 1 yard plus a half a yard plus 1/3 [yard]. And then what I did was I turned this into pies to show 1/2 and 1/3. I split this one in half. I split this one into thirds. And then I knew that they were getting 1/2 already so this is 1 1/2. And then I broke it down. I couldn't add the 1/2 and 1/3 so I broke this down into sixths. I cut them down both and I know that there were 1, 2, 3 sixths. And then I had 2 more sixths. And if I add them altogether I get 5/6. So that's 1 yard and 5/6.

Instructor: Questions for Lilly? David.

David: How did you get the sixth? Why did you divide it into six?

Lilly: Because I couldn't add these two numbers. So I broke it down into something where I could figure out how many were in a half and how many were in each third. So I broke down both pies into six.

Matt: I think we have to explain it better than that. Like why.

The classroom dialogue continued with students working together in an attempt to provide an acceptable explanation to Lilly's solution according to the agreed upon criteria for what counts as an acceptable explanation detailed in the previous section. By requesting this information from Lilly, Matt was taking an active role in sustaining accepted norms. This episode is representative of many classroom interactions in which the students actively sustained norms and the instructor's function was limited to that of a supporting role.

Conclusion

The importance of establishing social and sociomathematical norms is well documented in the literature (Cobb et al., 1992; Stephan & Whitenack, 2003; Yackel & Cobb, 1996; Yackel et al., 2000); however, relatively few studies have shown what norms are established among college age students, particularly pre-service teachers. In our study, we used the results that researchers found to be effective for developing mathematical reasoning in elementary and undergraduate mathematics classrooms to support the development of an inquiry environment in an undergraduate elementary mathematics course for prospective teachers. This study sought to add to the literature by making transparent the process of establishing and maintaining these norms with prospective teachers. We found that a teacher, and in our case the research team, that successfully establishes norms in an inquiry environment must be pro-active by creating opportunities that allow for norm negotiation in her classroom, not wait for norms to emerge on their own. We also noted that, while it is crucial to plan opportunities for norms to arise in classroom discussions, the teacher must be able to adapt to the *actual* negotiations that take place, recognizing that they may not go as planned. The implication is that a teacher must understand and value the norm building process so that she may recognize fruitful opportunities and capitalize on them at the moment. Finally, a successful teacher knows that norms cannot be built in one class period, and although they may appear fairly stable at any point in time, she must continue to reflect on classroom discourse so that she can sustain positive norms and recognize when new norms arise or current norms become unstable.

Our approach to initiating and sustaining social and sociomathematical norms was developed through the supportive nature of a classroom teaching experiment in which the instructor was not the sole decision-maker nor the sole expert observer of interactions in the classroom. Research needs to be conducted to apply this approach in settings outside the classroom teaching experiment where a research team is neither practical nor available.

References

Andreasen, J. B. (2006). Classroom mathematical practices in a preservice elementary mathematics education course using an instructional sequence related to place value and operations. Unpublished Dissertation, University of Central Florida, Orlando, FL.

Cobb, P. (2000). Conducting teaching experiments in collaboration with teachers. In A. E. Kelly & R. A. Lesh (Eds.), Handbook of research design in mathematics and science education (pp. 307–333). Mahwah, NJ: Lawrence Erlbaum Associates, Inc.

Cobb, P., & Whitenack, J. W. (1996). A method for conducting longitudinal analyses of classroom videorecordings and transcripts. Educational Studies in Mathematics, 30, 213–228.

Cobb, P., Wood, T., Yackel, E., & McNeal, B. (1992). Characteristics of classroom mathematics traditions: An interactional analysis. American Educational Research Journal, 29(3), 573–604.

Glaser, B. G., & Strauss, A. L. (1967). The discovery of grounded theory: Strategies for qualitative research. New York: Aldine De Gruyter.

National Council of Teachers of Mathematics. (2000). Principles and standards for school mathematics. Reston, VA: Author

Simon, M. A. (2000). Research on the development of mathematics teachers: The teacher development experiment. In A. E. Kelly & R. A. Lesh (Eds.), Handbook of research design in mathematics and science education (pp. 335–359). Mahwah, NJ: Lawrence Erlbaum Associates, Inc.

Steffe, L. P., & Thompson, P. W. (2000). Teaching experiment methodology: Underlying principles and essential elements. In A. E. Kelly & R. A. Lesh (Eds.), Handbook of research design in mathematics and science education (pp. 267–306). Mahwah, NJ: Lawrence Erlbaum Associates, Inc.

Stephan, M., & Whitenack, J. W. (2003). Establishing classroom social and sociomathematical norms for problem solving. In F. K. Lester & R. I. Charles (Eds.), Teaching mathematics through problem solving: Prekindergarten — grade 6. Reston, VA: National Council of Teachers of Mathematics.

Wheeldon, D. (2008). Developing mathematical practices in a social context: An instructional sequence to support prospective elementary teachers' learning of fractions. Unpublished Dissertation, University of Central Florida, Orlando, FL.

Yackel, E. (2001). Explanation, justification, and argumentation in mathematics classrooms. Paper presented at the International Group for the Psychology of Mathematics Education, Utrecht, The Netherlands.

Yackel, E., & Cobb, P. (1996). Sociomathematical norms, argumentation, and autonomy in mathematics. Journal for Research in Mathematics Education, 27, 458–477.

Yackel, E., Rasmussen, C., & King, K. (2000). Social and sociomathematical norms in an advanced undergraduate mathematics course. Journal of Mathematical Behavior, 19, 275–287.

Juli K. Dixon is a professor of mathematics education at the University of Central Florida. Her research interests relate to communicating and justifying mathematical ideas and developing and deepening prospective and in-service teachers' mathematics content knowledge for teaching.

Janet B. Andreasen is a visiting assistant professor at the University of Central Florida. Her research interests include the development of mathematical knowledge for teaching at elementary and secondary levels as well as the use of technology in preparing prospective teachers for the classroom.

Michelle Stephan is a full-time 7th grade teacher at Lawton Chiles Middle School and an associate graduate faculty member at the University of Central Florida. Her research interests include creating innovative instructional sequences for middle school using the Realistic Mathematics Education instructional design theory. She is also interested in the role that the teacher plays in creating and sustaining inquiry-based learning environments in K–12 settings.

Markworth, K., Goodwin, T. and Glisson, K.
AMTE Monograph 6
Scholarly Practices and Inquiry in the Preparation of Mathematics Teachers
© 2009, pp. 67–83

5

The Development of Mathematical Knowledge for Teaching in the Student Teaching Practicum

Kimberly Markworth
University of North Carolina, Chapel Hill

Tijuana Goodwin
Chapel Hill–Carrboro City Schools

Kelsey Glisson
University of St. Andrews

Using a case study approach we investigated the relationship between a student teacher and cooperating teacher to answer the following questions: What mathematical knowledge for teaching is acquired from cooperating teachers in the student teaching practicum? How does this learning take place? The findings suggest that the student teaching practicum is a rich site for the development of mathematical knowledge for teaching, particularly in three domains. This knowledge was acquired through practical experience with teaching, modeling, discourse with and observation of the cooperating and other teachers, analysis of student work, and studying textbooks. Implications for practicum experiences in pre-service education and future research are discussed.

The student teaching practicum is a fundamental aspect of most undergraduate teacher preparation programs. The format of the experience varies, but each pre-service teacher is usually paired with a veteran teacher for a semester-long internship. During this time, the student teacher observes the cooperating teacher, gradually picks up duties, and spends several weeks assuming full teaching responsibilities. Students often cite student teaching as the most beneficial aspect of their programs to their preparation for the teaching profession (Feiman-Nemser, 1983; Feiman-Nemser & Buchmann, 1989).

However, there is remarkably little research investigating what student teachers learn from their cooperating teachers. Veteran teachers who mentor student teachers are a vast source of knowledge about their subject matter and how to teach it. In this immersive setting, student teachers acquire content knowledge, knowledge about students, teaching methods, classroom management, and more. In this study we utilized a case study approach to investigate the relationship between one student teacher (third author) and her cooperating teacher (second author). We attempted to deepen understanding of the following questions: What mathematical knowledge for teaching is acquired from cooperating teachers in the student teaching practicum? How does this learning take place?

Mathematical Knowledge for Teaching

Shulman's presidential address to the annual conference of the American Educational Research Association in 1985 was monumental in its impact on the educational research community. His address and its subsequent publication in *Educational Researcher* (Shulman, 1986) coined the term *pedagogical content knowledge* (PCK) as knowledge "which goes beyond knowledge of subject matter per se to the dimension of subject matter knowledge *for teaching*" (p. 9). This new construct went beyond domain-specific content knowledge and knowledge of general pedagogy to include

> the ways of representing and formulating the subject that make it comprehensible to others…, an understanding of what makes the learning of specific topics easy or

difficult…, and the knowledge of the strategies most likely
to be fruitful in reorganizing the understanding of learners.
(p. 9–10)

Since 1986, an ever-growing body of research has sought to
develop this construct in various content areas.

Through fifteen years of research on knowledge related to
the teaching of mathematics, researchers at the University of
Michigan have identified six domains of mathematics knowledge
for teaching (Ball, Thames, & Phelps, 2008). Three domains,
common content knowledge (CCK), *specialized content
knowledge* (SCK), and *horizon knowledge*, are categorized as
subject matter knowledge. The other three domains, *knowledge
of content and students (KCS)*, *knowledge of content and*
teaching (KCT), and *knowledge of content and curriculum*, are
categorized as pedagogical content knowledge. Three domains
will be highlighted for the present study: specialized content
knowledge; knowledge of content and students; and knowledge
of content and teaching.

Specialized content knowledge is content knowledge needed
for the teaching of mathematics, beyond the common content
knowledge needed by others. For example, although many
people know how to carry out a subtraction problem involving
regrouping and recognize a correct or an incorrect answer
(CCK), teachers' knowledge must go beyond that. They must not
only be able to identify an incorrect answer, but also the methods
and the faulty reasoning which were used to produce it (SCK).
This understanding is entirely mathematical; it requires neither
knowledge of students nor knowledge of teaching. The domains
under the umbrella of pedagogical content knowledge, however,
combine knowledge of mathematics with knowledge of students,
teaching, and curriculum. Knowledge of content and students
encompasses the anticipation of the mistakes and
misunderstandings students might exhibit in relation to particular
content. Knowledge of content and teaching includes knowledge
of how the mathematics instruction should be designed and
sequenced, how and when students' responses should be drawn
upon, and benefits of particular mathematical representations.

Ball et al. (2008) admitted that further refinement of these categories of mathematical knowledge for teaching may be necessary. They also noted that the categories are static and do not reflect their use in practice and that the boundaries between categories are sometimes fuzzy. Ball et al (2008) further stated:

> The shifts that occur across the four domains, for example, ordering a list of decimals (CCK), generating a list to be ordered that would reveal key mathematical issues (SCK), recognizing which decimals would cause students the most difficulty (KCS), and deciding what to do about their difficulties (KCT), are important yet subtle. (p. 404)

The deficit of mathematical knowledge for teaching in both pre-service and practicing teachers is well-documented (see, for example Ball, 1990; Ma, 1999). Additional studies assess teachers' mathematical knowledge for teaching using a comparative approach and suggest that this knowledge may develop with time and experience teaching. For example, Leinhardt and Smith (1985) compared experienced and novice teachers' mathematical knowledge for teaching fractions and found that knowledge grows with experience. Similarly, Borko and Livingston (1989) compared three student teachers and their cooperating teachers with respect to their thinking and actions in teaching mathematics. They found that as teachers become more experienced, they are better able to anticipate and address students' misconceptions and questions. Fuller's (1997) survey of novice and experienced elementary teachers indicated that experienced teachers surpassed novice teachers in their conceptual understanding of whole number operations. However, both groups demonstrated predominantly procedural understanding of fractions, which is inconsistent with the results of studies cited above that showed that mathematical knowledge develops through experience.

Because most pre-service teachers participate in student teaching with a veteran teacher, this experiential site provides opportunities for veteran teachers to share and pre-service teachers to develop mathematical knowledge for teaching. Peterson and Williams (2008) used contrasting cases to compare the core conversational themes discussed by two cooperating teacher/student teacher pairs. One pair focused its discussions predominantly on classroom management while the other pair focused on sixth-grade mathematics content and students' active participation in the classroom. These conversations reflected the cooperating teachers' differing beliefs about the teaching of mathematics with the former focused on "students' compliance" and the latter focused on "students' understanding" (p. 476). In one case, mathematical knowledge for teaching was successfully developed in the practicum experience, but in the other case, the learning was limited.

To conduct an initial exploration into what and how mathematical knowledge for teaching is developed during student teaching, the first author of this chapter conducted a study of the exchanges between the second and third authors as they functioned as a cooperating teacher/student teacher pair. Because little research has been conducted on how mathematical knowledge for teaching (MKT) is developed, and only one study of MKT development has been conducted during the student teaching practicum (Peterson & Williams, 2008), the two questions guiding this research are exploratory in nature. Piotrkowski (1979) posits that "small-sample research and qualitative techniques are especially appropriate to research problems that require exploration and the discovery of patterns, hypotheses, and theory rather than their validation" (p. 291). Qualitative case study methodology was used for this study to allow for inductive reasoning about the domains of MKT, what knowledge is developed through experience, and how it is developed.

Methods

During the fall semester at the university, student teachers in the middle grades program are assigned to their practicum placement. Each week, the student teacher spends three hours in this classroom getting to know the teacher's routines and teaching style. This study was conducted in the spring semester while the student teachers were in their placements full-time through April, gradually assuming teaching responsibilities.

Participants

Purposeful sampling was used to select the student and cooperating teacher for this study. One student teacher enrolled in the final semester of the middle grades mathematics program was selected based on her demonstrated thoughtfulness with mathematics content and likely willingness to reflect openly on her practice. Similar characteristics of her cooperating teacher were also considered. At the time, the cooperating teacher taught two pre-algebra classes and two Algebra I classes in the eighth grade. Both the cooperating and student teachers are representative of veteran and pre-service teachers working with the middle grades mathematics program at the university in terms of demographics, experience in the classroom, and experience with mathematics and pedagogy coursework.

Data Collection Strategies

Three different forms of data collection were used in this case study to allow for triangulation of data: participant observation, interviews, and document collection. Participant observation occurred on a weekly basis for five consecutive weeks during the data collection time period. Each week, the first author observed one mathematics class taught by the student teacher and took field notes to record the general progression of the lesson. After the classroom observation (usually the following day), a reflective conversation between the cooperating teacher and student teacher was observed and audio-recorded and later transcribed. Although such conversations are common, these were likely unusual because they were compulsory (for research purposes) and the teachers were directed to reflect on the lesson from the previous day, discussing whatever topics they felt were important. Generally, the researcher did not participate actively in these conversations.

Each participant was interviewed individually twice–once at the beginning of the student teaching practicum and once at the end, and all interviews were audio-recorded and transcribed. The first interview focused on what the teachers expected to learn from each other and how they hoped to teach or develop knowledge throughout the practicum experience. The second interview focused on the participants' perceptions of what had been learned or taught during the practicum experience as well as how it was learned or taught. Four mathematics questions, modeled after sample items released by Hill, Schilling, and Ball (2004), were written to address the teachers' knowledge of polynomials and were used as part of each interview (see Appendix A for two of these questions). Polynomials were chosen because the student teacher taught a unit on polynomials to the Algebra I students during her practicum.

A student teacher/cooperating teacher notebook was kept in which the cooperating teacher made notes as she observed her student teacher. The notebook was used primarily on days when research observations took place, and the notes therein served mainly to guide post-lesson conversations between the cooperating teacher and student teacher.

Data Analysis

Transcripts of interviews were used to summarize answers to all interview questions. Likewise, transcripts of post-lesson conversations were summarized into main topics. Each conversation spanned 11 to 16 topics, with some topics containing several subtopics related to the main topic. Interview responses and conversational topics were then coded based on a priori themes according to three domains of mathematical knowledge for teaching: *specialized content knowledge, knowledge of content and students,* and *knowledge of content and teaching.*[i] As noted previously, the boundaries between these domains can be fuzzy; in some cases, responses and topics were coded for primary, secondary, and even tertiary domains. In the Findings section, the cooperating teacher is referred to as CT and the student teacher as ST in transcripts.

Findings

> I think as far as just basic content knowledge—her unit
> is going to be on factoring and polynomials and things of
> that sort—I think basic mathematical knowledge she has
> down. I think it is figuring out how to take that
> knowledge and teach it in such a way to the kids so that
> they understand what she knows. (CT, First Interview)

This comment from the beginning of the student teaching
practicum suggests that the cooperating teacher had a limited
view of what mathematical knowledge the student teacher would
develop during her experience. Instead of mathematics content,
the cooperating teacher expected the student teacher to learn how
to teach effectively her already-acquired mathematical
knowledge. Indeed, the student teacher's mathematical
knowledge for teaching grew in many ways. The findings below
are presented according to the three domains of mathematical
knowledge for teaching: specialized content knowledge;
knowledge of content and students; and knowledge of content
and teaching.

Specialized Content Knowledge
The student teacher's understanding of factoring
polynomials was enhanced through teaching the content, which
contributed to her specialized content knowledge. In particular,
she learned a more efficient method for factoring trinomials,
called "the box method," which was referred to repeatedly
during the interviews and conversations.[ii] The student teacher
found this method especially helpful for factoring trinomials in
which the first coefficient in standard form was a number other
than one. As she reflected on her own learning of factoring
polynomials in comparison to this new method, the student
teacher noted, "Well, I guess what I learned most of all was I
was never taught using a box method of any kind. So I learned
by trial and error: maybe this will work, maybe it won't.... I
think that using the box method is the most efficient way of
teaching [factoring trinomials]" (ST, Second Reflection).

Because this factoring procedure is knowledge that she learned specifically for teaching, it is therefore part of her specialized content knowledge.

In addition, the student teacher demonstrated knowledge growth in her responses to the mathematical items in the interviews. For example, in her first interview, she did not recognize that a LOFI (a modified version of the FOIL mnemonic) would work for multiplying two binomials (Question 2, Appendix A). During the second interview, the student teacher worked out a hypothetical problem using LOFI and recognized its mathematical accuracy. She identified the reasoning behind this solution strategy in order to judge the accuracy of the strategy, which falls under the domain of specialized content knowledge. She also indicated that the student should be allowed to use any method that both worked mathematically and was comfortable for the student.

The cooperating teacher also indicated that the student teacher became more knowledgeable of vocabulary relating to polynomials. Specifically, she came to a more refined understanding of the terms monomial, binomial, trinomial, and polynomial. This is knowledge that is necessary for a teacher in order to produce coherent and consistent presentations; fluency with terminology is not necessary for the average layperson. This knowledge, like the box method for factoring trinomials, is specialized content knowledge that the cooperating teacher did not necessarily expect the student teacher to learn.

Knowledge of Content and Students

In her second interview, the cooperating teacher identified aspects of knowledge of content and students that the student teacher had learned during the semester, including common student misconceptions and errors. When asked how her student teacher's knowledge of polynomials had grown through teaching the content, the cooperating teacher responded:

> How students confuse, like if you have x and x^2, how they still somehow, no matter how much you say to them over and over and over and over again, even with

her Lab Gear [a manipulative], this is an x, this is an x^2,
they would still get that concept confused. So I think her
just learning to be patient and also just to repeat things
that she's said, and also just trying to get at how kids
thought about the mathematics so that she can somehow
get in there and try to help them deal with the
misconceptions that they had.... Some kids, she realized,
were confusing addition rules and multiplication rules.
So, if I have an x^2 plus an x^2, dealing with that versus
an x^2 times an x^2. (CT, Second Interview)

These common student misconceptions surfaced repeatedly in
conversation, as the cooperating and student teacher discussed
students' errors or misapplications of previously learned
concepts and algorithms. The student teacher found that she
needed to revisit content so that she could address these
misconceptions.

The student teacher learned that there are common
misconceptions and errors that can be anticipated when tackling
this particular content with students. In the future, this
knowledge will help her respond quickly to these errors and
explore alternatives to teaching the content in order to minimize
student misconceptions.

Knowledge of Content and Teaching
    The benefits of using manipulatives and different classroom
activities are aspects of knowledge of content and teaching. The
student teacher's use of Algebra Lab Gear throughout her unit on
polynomials enabled her to see how the use of a concrete
representation of concepts enhances students' conceptual
understanding. She stated, "Using manipulatives is very helpful.
I've always been told that [about] using manipulatives, but
learning firsthand how helpful that can be for establishing
conceptual understanding is a big [thing that I learned]" (ST,
Second Interview). However, she also recognized their
limitations, noting in a reflective conversation that although

Algebra Lab Gear could represent x and x^2 effectively, it could not show the product of $x^2 \cdot x^2$ in a concrete way.

Growth in knowledge of content and teaching was also evident in her responses to the third mathematical interview question (Question 1, Appendix A). In the first interview, prior to teaching her unit on polynomials, the student teacher did not see how the garden plot/area model representation could be extended to demonstrate multiplication of two binomials. Following her unit, during which she used Algebra Lab Gear and the box method for factoring trinomials, the student teacher immediately was able to see how the representation could be extended and did so by continuing the left side of the rectangle to represent $g + 7$.

Developing Mathematical Knowledge for Teaching

The student teacher and cooperating teacher identified different sources of the student teacher's growth of mathematical knowledge for teaching. The student teacher attributed her learning to gaining teaching experience, observing and communicating with other teachers, studying mathematics textbooks, and discussions with her cooperating teacher. For example, the student teacher first encountered the box method in her study of an algebra textbook. However, her understanding of the method was enhanced when she solved some problems and discussed its use with the cooperating teacher. At the beginning of the semester, the student teacher did not expect to learn in these ways. Rather, during her first interview, the student teacher indicated that she only expected to learn through observation and trial and error.

In contrast, the cooperating teacher anticipated at the beginning of the study that the student teacher would learn through interaction with other teachers. The cooperating teacher also attributed the student teacher's learning to the cooperating teacher's modeling and demonstration of alternate methods as well as the student teacher's close analysis of student work. Although the cooperating teacher indicated that she contributed to the student teacher's learning through modeling and discourse, she attributed most of the learning to the practical experience of teaching.

The analysis of the post-lesson conversations demonstrates how discourse contributed to the student teacher's learning in three domains of mathematical knowledge for teaching. Although one might expect much conversation to revolve around classroom management (Feiman-Nemser, 1983), this was not the case. The majority of each conversation was mathematical in nature. Within the domain of specialized content knowledge, the cooperating and student teacher discussed terminology and mathematical connections, mathematical properties, and alternatives to traditional algorithms, as with the box method for factoring trinomials. In the domain of knowledge of content and students, the pair most often discussed students' misconceptions, questions, and comments encountered during the lesson, as well as the origin of the students' confusions. Often, the cooperating teacher followed these brief segments with suggestions and options for teaching particular content. Thus, the knowledge of content and students that was developed overlapped with knowledge of content and teaching. By discussing how best to address misconceptions, questions, and comments, the pair linked student understanding to instructional practice. Other recommendations for teaching practice were made by the cooperating teacher spontaneously (i.e., not in response to a concern raised by the student teacher). For example, in planning for future lessons, the cooperating teacher suggested that the student teacher approach special cases (i.e., factoring $4x^2 - 9$) by letting the students grapple with the unusual situation.

Discussion

This exploratory research demonstrates that the student teaching practicum experience is a potentially rich site for developing mathematical knowledge for teaching. The student teacher in this case study gained specialized content knowledge, knowledge of content and students, and knowledge of content and teaching. In particular, the student teacher learned subject matter knowledge about polynomials that included a more refined understanding of vocabulary and a new method for factoring trinomials. Her pedagogical content knowledge was augmented by learning about common student errors and misconceptions, the effective use of manipulatives, and useful representations of the subject matter.

Although discourse with and observation of the cooperating teacher were critical to the student teacher's growth, she also reported learning by studying texts and through other teachers in the school setting. Perhaps not surprisingly, both the student teacher and cooperating teacher attributed a lot of the learning to the practical experience of teaching.

Efforts should be made in pre-service teacher education to capitalize on the practicum experience as a site for developing domains of mathematical knowledge for teaching. One way to accomplish this may be through work with cooperating teachers. This particular cooperating teacher did not anticipate her student teacher learning mathematical knowledge; often student teachers' knowledge of content is assumed to be sufficient for teaching. Veteran teachers have acquired knowledge for teaching mathematics through experience and may be unaware of this experience-based knowledge. If cooperating teachers were helped to become aware of their deep mathematical knowledge for teaching, they might be more apt to make it explicit and encourage its development in their student teachers.

Additionally, during the practicum experience, efforts should be made to foster and encourage reflective mathematical post-lesson discourse. Time is often scarce for critical, prolonged reflection on student teachers' lessons. Although time-consuming, the weekly reflection that this research afforded was appreciated, especially by the student teacher. She stated in her final interview:

> This research project has... enabled us and made us sit down and talk about things, where we might not necessarily have talked as in depth about the things that we've talked about with you. So, this has taught me a lot about reflection and inquiry and making sure you reflect on what you teach so that you can improve on it later.

Her comment displays the importance of reflection and inquiry and its potential for helping student teachers develop mathematical knowledge for teaching when the conversation is focused on mathematics. Further, it is important for pre-service teachers to be aware of the vast amount of learning that they still have ahead of them as they enter the profession. If they are reflective practitioners, open to the possibility of career-long learning about mathematics, their students, and their teaching, they will continue to develop and deepen their mathematical knowledge for teaching.

It is our hope that this exploration identifies future possibilities for research. This may include experimental research on efforts to enhance the development of mathematical knowledge for teaching during the practicum experience. Or, it may include a more in-depth exploratory analysis of the other avenues for this knowledge development. There is certainly more to be learned from student teachers' work with planning instruction, discourse with and observations of other teachers, and modeling of cooperating teachers' practices. Despite its prevalence in pre-service teacher education programs, the practicum experience remains largely unexplored territory. However, it is a rich site for learning, and research on this experience will enhance teacher educators' knowledge of how this experience can be optimized.

References

Ball, D. L. (1990). The mathematical understandings that prospective teachers bring to teacher education. *The Elementary School Journal, 90*(4), 449–466.

Ball, D. L., Thames, M. H., & Phelps, G. (2008). Content knowledge for teaching: What makes it special? *Journal of Teacher Education, 59*(5), 389–407.

Borko, H., & Livingston, C. (1989). Cognition and improvisation: Differences in mathematics instruction by expert and novice teachers. *American Educational Research Journal, 26*(4), 473–498.

Feiman-Nemser, S. (1983). Learning to teach. In L. S. Shulman & G. Sykes (Eds.), *Handbook of teaching and policy* (pp. 150–170). New York, NY: Longman, Inc.

Feiman-Nemser, S., & Buchmann, M. (1989). Describing teacher education: A framework and illustrative findings from a longitudinal study of six students. *The Elementary School Journal, 89*(3), 365–377.

Fuller, R. A. (1997). Elementary teachers' pedagogical content knowledge of mathematics. *Mid-Western Educational Researcher, 10*(2), 9–16.

Hill, H. C., Schilling, S. G., & Ball, D. L. (2004). Developing measures of teachers' mathematics knowledge for teaching. *The Elementary School Journal, 105*(1), 11–30.

Leinhardt, G., & Smith, D. A. (1985). Expertise in mathematics instruction: Subject matter knowledge. *Journal of Educational Psychology, 77*(3), 247–271.

Ma, L. (1999). *Knowing and teaching elementary mathematics: Teachers' understanding of fundamental mathematics in China and the United States.* Mahwah, NJ: Erlbaum.

Murdock, J., Kamischke, E., & Kamischke, E. (2007). *Discovering algebra: An investigative approach* (2nd ed.). Emeryville, CA: Key Curriculum Press.

Peterson, B. E., & Williams, S. R. (2008). Learning mathematics for teaching in the student teaching experience: two contrasting cases. *Journal of Mathematics Teacher Education, 11*, 459–478.

Piotrkowski, C. S. (1979). *Work and the family system.* New York, NY: The Free Press.

Shulman, L. S. (1986). Those who understand: Knowledge growth in teaching. *Educational Researcher, 15*(2), 4–14.

Appendix A

1. Last year in pre-Algebra, students learned the distributive property by finding the areas of rectangular garden plots.

The following representation demonstrates that $g^2 + 3g$ is equivalent to the expression $g(g + 3)$.

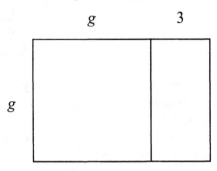

How could you build on this representation to show your students how to multiply two binomials, such as $(g+3)(g+7)$?

2. Louise, a student in your class, has just been shown how to multiply two binomials using the FOIL algorithm (First, Outer, Inner, Last). She asks, "Do we have to use FOIL? Won't we get the same answer if we do LOFI?" How would you respond to her question?

Endnotes

[i] Coding occurred along four domains of MKT: common content knowledge; specialized content knowledge; knowledge of content and students; and knowledge of content and teaching. However, since the mathematics that was developed was learned through and for the practice of teaching, nothing was primarily coded as common content knowledge.

[ii] The "box method" is based on an area model and referred to as a rectangle diagram in one algebra text (Murdock, Kamischke, & Kamischke, 2007).

Kim Markworth is a doctoral student at the University of North Carolina at Chapel Hill, and taught elementary school and middle grades mathematics for 12 years. Her research interests include mathematical knowledge for teaching, preservice teacher education, and the teaching and learning of algebra.

Tijuana Goodwin has been teaching for 15 years, and currently teaches 6th, 7th, and 8th grade mathematics at McDougle Middle School in Chapel Hill, North Carolina. Some of her interests include mathematics education, literacy, and personal growth and development.

Kelsey Glisson is a Master's student at the University of St. Andrews in St. Andrews, Scotland, where she is pursuing a Master's in Finance and Management. Her interests include mathematics education, business management, and marketing.

Rathouz, M. and Rubenstein, R. N.
AMTE Monograph 6
Scholarly Practices and Inquiry in the Preparation of Mathematics Teachers
© 2009, pp. 85–103

6

Supporting Preservice Teachers' Learning: A Fraction Operations Task and its Orchestration

Margaret Rathouz
Rheta N. Rubenstein
University of Michigan-Dearborn

As teacher educators, we strive to establish learning experiences for pre-service teachers (PSTs) that support them in their transition from a rules-based to a reasoning-based orientation to mathematics learning. In doing so, we consider the roles of a robust mathematical task and its classroom implementation. As we study video records of a learning community and its interaction with the task, we seek insights into the strategies that PSTs use to untangle their confusions and justify their mathematical claims. The analysis presented here provides the reader a glimpse of efforts to build an inquiry classroom for pre-service elementary teachers, with the aim that these future teachers will one day teach in comparable ways.

To teach mathematics in classrooms where children learn to reason and justify their thinking, future teachers benefit by experiencing a comparable environment of inquiry themselves (National Council of Teachers of Mathematics (NCTM), 1991; Conference Board of Mathematical Sciences (CBMS), 2001). As teacher educators, we strive to establish such learning experiences for pre-service teachers (PSTs) and to support them in their transition from an orientation to mathematics that is rules-based toward one based on reasoning. In doing so, we

consider the roles of a robust mathematical task and its
classroom implementation (Stein, Smith, Henningsen, & Silver,
2000; Yackel & Cobb, 1996).

The research of Stein et al. (2000) reminds us that the
selection of mathematical tasks and their enactment in ways that
maintain their cognitive demand are fundamental to providing
K–12 students opportunities to learn mathematics with deep
understanding. However, there has been little research
identifying how features of a task and its implementation provide
appropriate challenge for PSTs. To fill this gap in the literature,
we analyze elements of one task and the discussion it initiates in
a mathematics content course for PSTs. Our goals are to better
understand PSTs' thinking about fraction operations and the
ways mathematics educators may deepen that understanding.

Our work in developing a mathematics curriculum for PSTs
is a collaborative effort between several mathematics educators
on two college campuses. One of these colleagues (who is not a
co-author) was the instructor for the course from which
transcripts here are shared[1]. The authors of this paper have taught
the course and are involved in its on-going development and
analysis. This paper intends to add to a growing body of studies
of selected tasks from our curriculum (Flowers, Kline, &
Rubenstein, 2003; Flowers, Krebs, & Rubenstein, 2006; Flowers
& Rubenstein, 2006; Grant, Lo, & Flowers, 2007; Lo, Grant, &
Flowers, 2008; Rathouz, in press).

Hiebert and Grouws (2007) suggested that detailed analyses
are needed in order to draw meaningful conclusions about
teaching and its relationship to learning. To that end we have
collected video records of our first course in mathematics
content for PSTs. As we study these records and reflect on our
teaching, we seek insights into the following questions. What
strategies do adult students use to untangle their confusions?
How do they justify their claims? How do the tasks themselves
and the instructors' implementations help to promote a
mathematical community of inquiry and justification? We invite
readers to consider these questions as we share this sample of
instruction.

Introducing the Task

The instructional segments we examine utilize a fraction task from the second unit in a first mathematics content course for elementary PSTs. While the type of task is well known to many educators, we believe studying PSTs' thinking around it and the instructor's responses help us better understand and support PSTs' learning. Throughout the paper, we refer to the PSTs as "students" and the mathematics educator as the "instructor" (labeled "T" in the transcripts). Ellipses (...) indicate the omission of a small segment of the transcript. The first unit in the course had focused on whole numbers, and students were encouraged to rely on meanings of operations to justify solutions to arithmetic problems. Just prior to the instruction described here, initial work with fractions had emphasized the importance of the referent whole. At this point in the semester, the students were asked to generate story problem situations for subtraction of fractional numbers. Below, the task of writing story problems for $3\frac{1}{2} - \frac{2}{3}$ was introduced by the instructor.

> T: To help us think more about subtraction, I want you to work together in your groups to come up with a story problem for that: [T points to "$3\frac{1}{2} - \frac{2}{3} = ?$" written on board.] Not the solution....don't figure out the answer...just a story for $3\frac{1}{2} - \frac{2}{3}$. I want you to work in your groups to come up with at least two different kinds of stories for that problem.

Notice that the focus is not on calculation but rather on generating stories that "feel" different. The reader is encouraged to attempt the task before reading on and to consider its goals, the mathematical reasoning it may bring to the fore, and approaches pre-service teachers might take.

Students' Initial Responses

Groups posted their story problems on the board, and the class chose to focus on a story where the group presenting was unsure which of two different wordings would be correct. The instructor allowed students to struggle, giving them the opportunity to explore and compare the two versions.

> T: So this group [wasn't] sure what to do, so they said, "John has to run three and a half miles. He took a break." They couldn't decide if they should say 'two-thirds of the way' or 'two-thirds of a mile.' [Then they asked] "How much further must he run?"

Two students offer their initial thoughts and rationale about which story problem matches the expression $3\frac{1}{2} - \frac{2}{3}$.

> S 1: It depends on whether you want to take it out of the whole or out of a single mile.

> S 2: In a way, you're looking at all three and a half miles and you're saying, here's the three and a half miles and... here we are on the number line and you're trying to figure out how much further you have to go to get to three and a half.

These two students are already revealing some of the important mathematics in the task and are suggesting strategies and representations that might be used to justify which interpretation they think is correct. The first student suggests that the difference in the two possible stories may depend somehow on the referent units. Despite this being significant, it is stated without elaboration, representation, or justification, so more discussion is needed. At this early point in the discussion, the second student does not build on the first student's remark. She does, however, refer to the notion of a number line and to a meaning of subtraction with her statement "how much further do you have to go?" Students are sent back briefly to their small groups to negotiate among themselves which wording to use and to produce a convincing argument for which one would match $3\frac{1}{2} - \frac{2}{3}$. By providing time for the entire group to think, discuss, and participate, the instructor facilitates a debate situation where students must justify their reasoning.

A Student Representation of "Two-Thirds of a Mile" and "Two-Thirds of the Way"

After providing small group time, the instructor asked students to discuss their current thinking. The excerpt below shows that she acknowledges that writing an appropriate story is challenging but also emphasizes that it is important work in building mathematical understanding and in preparing for their work as teachers. The instructor reminds students that the representations they create can be used as tools to help in both communicating their arguments and helping themselves and others clarify their mathematical thinking.

> T: OK, Time out. This is the difficult part for you all…trying to communicate to each other about your thinking…. The better you are able to articulate your thinking, the more comfortable you are…. As I went around to groups, most groups (not all) are convinced that one [version] is right and one is wrong. Most of you have pictures that you're somehow using to say which one's right or wrong…so I wondered if somebody would be willing to share their picture to get us started.

A student is volunteered by his peers to show and explain a diagram that his group generated (replication of the student diagram appears below). Although the student is tentative in his comments and does not commit to either story's interpretation, posting the representation is valuable in two ways. First, it creates a publicly shared visual centerpiece for the whole class discussion. Everyone can reflect on the two possible stories utilizing the same diagram. Second, because the number line was created by the students, they are responsible for answering questions about how it was created. The ensuing discussion reveals fragile areas in students' knowledge and illustrates their

attempts to help one another clarify their understandings of
fractions.

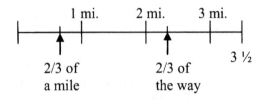

S 3:   …for this problem, I felt that even if your picture
       wasn't very much to scale, the amounts you're
       measuring are…so very different that it's…clear to see
       what you're measuring. You're measuring two-thirds
       of the way. 'The way' John is running is three and a
       half miles, so two-thirds of 'the way' would be a little
       bit over two miles 'til three and a half. [He points to
       the right arrow in the figure.] Two-thirds of 'a mile' …
       would be like right here [left arrow in the
       figure]….These are two, clear, different amounts.

S 4:   How did you get…two-thirds of the way?

S 3:   Two-thirds of the way, because if it's "of the way." I
       interpret that 'of'…as like "the three and a half miles."

S 4:   Then you find it for one mile first and then…

S 3:   Well, I just … estimated … I knew that two-thirds of
       three is two, so two-thirds of three and a half is gonna
       be a little bit more than two.

S 4:   Not exactly…

S 3:   No, but it's a picture, so it's not to scale. I just kinda
       made it up.

S 5:   Just to answer her question…you could've put like
       two extra marks in there between zero and the first

> mile and that would've broken your first mile into
> thirds, and you could've seen two-thirds.

Although this discussion has not answered the question of which wording, if either, matches the problem $3\frac{1}{2} - \frac{2}{3}$, progress is made on other fronts. The students are interacting without mediation from the instructor, taking responsibility for their own learning by asking clarifying questions (student 4), and attempting to elaborate on each others' thinking (student 5). Although the students find that both story problems are legitimate, the number line diagram helps to show that, depending on which interpretation is selected, a different answer results. This implies that non-equivalent mathematical expressions should be associated with each story. In the following summary of the discussion to this point, the instructor highlights Student 3's use of reasoning in his explanation, rephrases the original question, and extends the task by having students consider in their small groups what expression would fit the erroneous story.

> T: OK, this is progress, and, actually, I think you [St. 3]
>   [gave] a very nice explanation of how you knew two-
>   thirds of the way was more than two, because he knew
>   two-thirds of three was two, so if it's two-thirds of
>   three and a half, it's going to be a little more. So, that
>   was really nicely done. So, now, the 50 million dollar
>   question is, which one, if either of them, … matches
>   the problem $3\frac{1}{2} - \frac{2}{3}$, *and*, whichever one you decide
>   doesn't match $3\frac{1}{2} - \frac{2}{3}$, what [expression] *would* go
>   with it? So I'd like you to take a minute and chat with
>   each other about that.

Discussion of the Whole and the Unit

Having listened carefully to the small group deliberations, the instructor knows that arguments exist for each of the stories matching the original subtraction problem and relies on the strength of students' explanations to convince each other of their claim's validity.

T: ...I've been to the groups and some of them say it's "two-thirds of the way" and some of them say it's "two-thirds of a mile," and some of them are arguing within their groups.

S 7:   Two-thirds of the way is the right one. ...

[Lots of talking.]

T: [T interrupts heated discussion] Wait! ... Why do you say the "of the way" one ... matches this? [T points to original expression.]

S 7:   Because it takes it out of the whole, not out of a section.

T: So, she says it takes it out of the whole, not just [out of] a section of the whole. What do the rest of you say?

Sts [loudly]: Yeah! No!

Almost every student in the class is engaged in the debate. Student 7 seems to be saying in her last comment that the two-thirds cannot refer to just one mile because that would not take into account the original whole (three and a half miles). Although the student's reasoning is flawed, the instructor does not correct her. Instead, she asks other students to provide a counter-argument.

T: Somebody present the other argument for why it might be two-thirds of a mile.

S 8:   Two-thirds of a mile is keeping the same units, which is miles, and if it's two-thirds of three and a half miles, your unit's like three and a half miles total so it's...

[Many students speaking...]

T: Wait, hold on.

S 9:   It's keeping ... the same units as... two-thirds of a mile...I don't know how to say it...

S 10: No, you're right....keep going.

T: So, wait, you're saying three and a half miles minus two-thirds of a mile, you're keeping the units the same?

St 7: But "two-thirds of the way" you could say "two-thirds of the miles"

St 9: But two-thirds of the miles is two-thirds of three and a half.... [Lot's of talking]

T: OK, hold on. Did I record this right? So, hold on.... So, what I wrote...I was trying to write what you were saying and I think what he was saying was.... He sees the subtraction "three and a half miles minus two-thirds of a mile" versus the other one you see as "three and a half miles minus two-thirds of the three and a half miles." Did I write that right?

The previous segment begins to reveal students' conceptions of fractions and the operation of subtraction. Some of the issues are: What are we finding two-thirds *of*? What is the role of the whole (three and a half miles), the unit (one mile), and the part (two-thirds) in the expressions and the operations that go with each of the interpretations? Contributing to the struggle are issues related to how operations may or may not alter measurement units. It is unclear whether these PSTs know that one of the hallmarks of additive situations is that the units do not change, while in multiplicative contexts the units often do change. In their erroneous version of the story (where multiplication is implied), the units (miles) happen to remain constant because their two-thirds is a scalar. Arguments for either wording of the story problem need to appeal to more than just a discussion of the unit's <u>name</u> (the mile), but must also refer to its structure (two-thirds of a mile vs. two-thirds of three and a half miles). Student 9's initial comment attempts to clarify this distinction. The instructor captures, underscores, and highlights the state of the debate. She records for public reference their words relating the two stories in the diagram but refrains from giving away a symbolic interpretation.

Connecting Representations

Eventually, one student uses information about the story, the number line, and the symbolic representation to persuade herself and others that the "two-thirds of a mile" version makes sense.

S 11: ...if we do two-thirds of a mile and you figure out how far you've gone, even if you just approximate it, you could see that it wouldn't work if you did two-thirds of the way. Because, how much more do you have left? You have two and something, so you still have to go two more miles, but if we go two-thirds of the way, then you can see just from the loose drawing that we only have a mile and some to go....

S 12: And what are you saying is the right answer?

S 11: ...I think it's two-thirds of a mile because then you go two-thirds of a mile and then you subtract [and] $3\frac{1}{2} - \frac{2}{3}$ equals the rest of the way....

S 13: I can kinda see where you're coming from,...like I understand your argument and stuff. But, what I'm kinda confused on is like what are we looking at as the whole? Like you're saying we subtract three and a half from two-thirds (sic), that makes sense to me, but the problem is saying, "How much further must he run if he's at...three and a half miles, whoops...two-thirds of three and a half miles." Like one, you're kinda like changing it. It's like **one** is the whole.

T: We're definitely changing it, ... Just let me summarize an argument or two.... If you go two-thirds of the way, the question is how much is left? Right? Or wrong? So, if I subtract the two-thirds of the way, this is how much I have left to run. [T points to distance beyond right arrow.] And if I subtract two-thirds of a mile, this is how much is left to run. [T points to distance beyond left arrow.] OK, so,...so what was the essence of why she [St. 11] was saying this couldn't be right? Did you understand enough of her argument to repeat it? Can you understand it enough to rephrase it?

S 14   (from group that wrote the story): ... The reason we came up with...was the same as hers. So she's saying if you take $3\frac{1}{2} - \frac{2}{3}$ you're gonna have a larger number that, you know, a larger distance [is] left to go, so it has to be "of a mile"...

T: So the answer to this [$3\frac{1}{2} - \frac{2}{3}$] is about what? If you give me an estimate. ...

Sts: Two....

T: .... Do we all buy that? We know that the answer is two and something....So you're saying, you're convinced, that because of that [knowing the calculated answer is over two], that means that... [the version using "of a mile"] would have to be right?

Students 11 and 14 refer both to the approximate value of the calculation $3\frac{1}{2} - \frac{2}{3}$, a quantity that they know must be larger than two, and to the representations of the two versions for "How much further must he run?" on the number line, only one of which is larger than two miles. Although at the outset of the task the students were asked not to calculate a numeric answer, at this point several connections are made between the wordings of the stories, the referent units of the fractions, the number line representation, and the numeric approximation. The PSTs use these connections to strengthen their arguments and to convince more of their peers. We have found that if a calculation is used as the sole basis for a justification, the valuable confusion and resolution through discussion is short-circuited. There are still, however, some (e.g., St. 13) who are experiencing discomfort with the role that the "three and a half miles" should play in subtraction with fractions. ('Is three and a half miles the whole or is one mile the whole?'). The instructor helps to focus these skeptics by emphasizing the connections in her summary.

What Expression Goes with the "Of the Way" Story?
After more small group discussion, most of the students say they think the story that uses two-thirds of a mile (rather than two-thirds of the way) is the one that matches $3\frac{1}{2} - \frac{2}{3}$. As PSTs share their reasons for changing their minds, it is clear that students see a difference in the two versions of the story, although they note that both use subtraction. One of the goals of the fractions part of the course is to examine and analyze the distinction in wording of problems that involve solely fraction subtraction from those that involve subtraction with embedded multiplication. This piece of the task is explored in the final segment of the lesson. As the students begin to discover what expression could go with the story problem that did not match $3\frac{1}{2} - \frac{2}{3}$, ideas of fraction multiplication are introduced.

S 15: Now, I gotta know what the other one was [the calculation for the "of the way" story.]

T: Now you've got to know what that other one was. But I
    think he started us out in a nice place. He said it's three
    and a half miles minus two-thirds of three and a half
    miles....How do we write 'two-thirds of three and a half
    miles' as an expression? Without words. How do you
    write 'two-thirds of three and a half miles?'

In an attempt to connect the story problem to the symbolic
representation, the instructor refers to the words used earlier by
the students. To justify that two-thirds of three and a half miles
should be written as " $\frac{2}{3}$ x $3\frac{1}{2}$ ", the students must return to a
meaning of multiplication in a context where they were
comfortable. The students reflect on their work with whole
number multiplication using equal groups and begin to think
about two-thirds of three and a half miles as two-thirds of a
group of three and a half miles. Thus they make inroads into
understanding multiplication of fractions, which this lesson has
helped foreshadow. As is common with these lessons, the end of
one session stimulates thinking for upcoming work. (The next
lesson addressed multiplication more directly.) The homework,
too, asked students to revisit the ideas from class, including the
stories posted by the other groups, and, in this case, to clarify in
their own words the two types of stories that arose, their
representations, and symbolizations.

Analysis of the Learning Community

One of the striking features of this classroom is that thinking
is made public. Unlike rules-based classes where the instructor
provides explanations with examples that are replicated by
students, in reasoning-based classes, participants discuss ideas,
even those that are incomplete or misconstrued, to jointly
construct mathematical understanding. The work includes
learners building relationships among words, diagrams,
meanings, and symbols. Active debate makes clear the
expectation that reasoning is the basis for progress. More
importantly, the confusions and misconceptions held by many

students are not buried or lost; rather they are highlighted and incorporated as learning opportunities.

Actions of both the instructor and the students contributed to the learning community. As the facilitator, the instructor played many important roles (NCTM, 1991). For example, she posted several groups' story problems, asked a student to share his representation, and elicited students' interpretations. She listened closely to students and used their ways of thinking as the basis for the lesson development. She supported student thinking by building in wait time, encouraging further elaboration, and refocusing the students on parts of the task. She extended the mathematical thinking of the group by helping them make connections among representations, comparisons among operations, and generalizations from whole numbers to fractions. She created and took advantage of a teachable moment to begin to introduce meanings for fraction multiplication. In short, she elicited, supported, and extended students' thinking, moves that have been suggested by Fraivillig, Murphy, and Fuson (1999) to foster productive mathematical discourse.

The students also had many important roles (NCTM, 1991). They questioned each other during whole-class discussions and provided explanations and representations of their mathematical thinking. They responded to the class expectation of clarity and spontaneously built on their classmates' ideas. By questioning, explaining, and clarifying each others' thinking, the students accepted responsibility for their own learning and that of their peers, which are qualities of high-level math-talk communities (Hufferd-Ackles, Fuson & Sherin, 2004).

Analysis of the Role of the Task

A well-designed task is another critical element for a community's learning (Stein et al., 2000; NCTM, 1991), and this task has several features that make it worthwhile for PSTs to explore. First, it addresses multiple goals. On first pass, the reader may have envisioned that it would simply help PSTs match language to fraction subtraction, but the enactment of the task reveals its potential for addressing additional goals, many of which were mentioned in the transcript analysis and more of which appear below. Second, the task is relevant to PSTs in that it is situated in the practice of teaching; elementary teachers often write story problems to help students develop meanings for numbers and operations. But this mathematical knowledge for teaching is not trivial for PSTs.

A third feature is that the task "provokes disequilibrium" (Ball, 2008, p. 13). Asking students to "Write at least 2 different kinds of stories," nearly always draws out stories of both types $[3\frac{1}{2} - \frac{2}{3}$ and $3\frac{1}{2} - (\frac{2}{3} \times 3\frac{1}{2})]$. Deciding which story matches the original expression creates a struggle. Hiebert and Grouws (2007) describe struggling as "expending effort to figure something out that is not immediately apparent," "grappling with key mathematical ideas that are comprehendible but not yet well formed" or "novel problems…within reach, but that present enough challenge so that there is something new to figure out" (pp. 387–388). They argue that struggling promotes the development of mental connections among facts, ideas, and processes by putting the student in a place where he has no choice but to reconfigure his mental model for an idea. Further, the struggle in the task presented here involves a comparison, another feature that strengthens understanding (Star, 2008).

A fourth characteristic of the task that makes it valuable for PSTs is that it affords an opportunity for students to create and interpret representations. In this classroom the use of diagrams as thinking and communication tools is the norm. Diagrams produce a public referent and a tool for thinking, and they help PSTs build relationships among stories, number lines, and symbols.

Fifth, the task and its justification produce many common and mathematical language issues, which were illustrated when students discussed differences in wording. For example, two-thirds of "the miles" or two-thirds of "a mile" is a subtle change in language that has deeper mathematical implications for sorting out the difference between the unit and the whole. During the class period one PST asserted that "the word 'of' can't be in a subtraction problem." And another claimed that "'of' means 'times.'" In common language we might say, "take one-third of," as well as "take one-third off," where the former indicates solely multiplication without any subtraction and the latter uses similar language to indicate subtraction of one-third of some original quantity. We have found that PSTs need to confront these confusions in language. In doing so, they begin to see the need, ultimately, to analyze the relationships between the different representations and to build on *meanings* of the numbers and operations in order to develop convincing arguments.

In summary, the qualities of the story-writing task and the responses it elicited align with those of higher cognitive demand tasks characterized by Stein et al. (2000). The lesson required students to use non-algorithmic thinking; to represent their thinking in multiple ways; to explore and understand the nature of mathematical concepts, processes, or relationships; to access and use relevant knowledge and experiences; to examine constraints and limitations within a task; and to exert cognitive effort.

Closing Perspectives

This analysis provides the reader a glimpse of efforts to build an inquiry classroom. The teacher intends not only to support PSTs' mathematical learning, but also to bring to life for them our philosophy of reasoning-based learning. Our aim is that they might one day teach in comparable ways. We realize that they are at the beginning of a long growth curve, but these classroom experiences offer them a place to start.

We, too, are on a growth curve, learning to observe, understand, and think more critically about ways to support PSTs' mathematical understanding. One of our approaches, as illustrated in this paper, is to capture and examine noteworthy segments, analyze the instruction and tasks, and use insights gained for future planning and teaching. We hope that others may similarly learn from these analyses.

References

Ball, D. L. (2008, February). *The work of teaching and the challenge for teacher education.* http://www-personal.umich.edu/~dball/presentations/020908_AACTE.pdf. Presented at The American Association of

Colleges of Teacher Education Annual Meeting, New Orleans, LA.

Conference Board of the Mathematical Sciences (2001). *The mathematical education of teachers* (Vol. 11). Providence, RI: American Mathematical Society & Mathematical Association of America.

Flowers, J., Kline, K., & Rubenstein, R. N. (2003). Developing teachers' computational fluency: Examples in subtraction. *Teaching Children Mathematics, 9*(6), 330–334.

Flowers, J., Krebs, A., & Rubenstein, R. N. (2006). Problems to deepen teachers' mathematical understanding: Examples in multiplication. *Teaching Children Mathematics, 12*(9), 478–484.

Flowers, J., & Rubenstein, R. N. (2006). A rich problem and its potential for developing mathematical knowledge for teaching. In K. Lynch-Davis, R. L. Rider, and D. R. Thompson (Eds.), *The work of mathematics teacher educators: Continuing the conversation.* Monograph No. 3. San Diego, CA: Association of Mathematics Teacher Educators.

Fraivillig, J. L., Murphy, L. A., & Fuson, K. C. (1999). Advancing children's mathematical thinking in *Everyday Mathematics* classrooms. *Journal for Research in Mathematics Education, 30*(2), 148–170.

Grant, T. J., Lo, J., & Flowers, J. (2007). Shaping prospective teachers' justifications for computation: Challenges and opportunities. *Teaching Children Mathematics, 14*(1), 112–116.

Hiebert, J., & Grouws, D. (2007). The effects of classroom mathematics teaching on students' learning. In F. K. Lester, Jr. (Ed.) *Second handbook of research on mathematics teaching and learning* (pp. 371–404). Charlotte, NC: Information Age Publishing.

Hufferd-Ackles, K., Fuson, K.C. & Sherin, M. G. (2004). Describing levels and components of a math-talk learning community. *Journal for Research in Mathematics Education, 35*(2), 81–116.

Lo, J., Grant, T. J., & Flowers, J. (2008). Challenges in deepening prospective teachers' understanding of multiplication through justification. *Journal of Mathematics Teacher Education, 11*(1), 5–22.

National Council of Teachers of Mathematics (1991). *Professional standards for teaching mathematics.* Reston, VA: NCTM.

Rathouz, M. (in press). Supporting pre-service teachers' reasoning and justification: An example using related division problems. *Teaching Children Mathematics.*

Star, J. R. (2008). "It pays to compare! Using comparison to help build students' flexibility in mathematics." *The Center for Comprehensive School Reform and Improvement Newsletter.* Learning Point Associates. April 2008.

Stein, M. K., Smith, M. S., Henningsen, M. A., & Silver, E. A. (2000). *Implementing standards-based mathematics instruction: A casebook for professional development.* New York: Teachers College Press.

Yackel, E., & Cobb, P. (1996). Sociomathematical norms, argumentation, and autonomy in mathematics. *Journal for Research in Mathematics Education, 27*(4), 458–477.

Endnote

1. While this paper has been authored by two colleagues, several others have been involved in the instructional design, implementation, revision, and thinking related to this work. In particular, we are indebted to Dr. Theresa Grant of Western Michigan University for generously sharing her work and allowing the use of her videotapes. Major parts of this work were funded by the NSF DUE #0310829.

Margaret (Maggie) Rathouz is an assistant professor in the Department of Mathematics and Statistics at University of Michigan–Dearborn, in Dearborn, Michigan where she teaches courses for preservice and inservice teachers. She is interested in language, reasoning, and discourse in mathematics classrooms.

Rheta N. Rubenstein is a professor in the Department of Mathematics and Statistics at University of Michigan–Dearborn. She is interested in curriculum, instruction, and teacher education.

Gainsburg, J., Scheld, S., von Mayrhauser, C., Rothstein-Fisch, C. and Spagna, M.
AMTE Monograph 6
Scholarly Practices and Inquiry in the Preparation of Mathematics Teachers
© 2009, pp. 105–120

7

One Credential Program Looks in the Mirror

Julie Gainsburg, Suzanne Scheld, Christina von
Mayrhauser, Carrie Rothstein-Fisch, and Michael Spagna
California State University, Northridge

*This article describes the work of an interdepartmental
research team at California State University,
Northridge, to begin to systematically evaluate the
impact of its teaching credential program. First- and
second-year teachers who had graduated from our
secondary mathematics program were observed and
interviewed to determine the degree to which they
implemented program-emphasized teaching practices in
their classrooms and the factors—internal and external
to the program—that facilitated or impeded the
implementation of those practices.*

Introduction: *Teachers for a New Era*

In 2002, the Carnegie Corporation, with the Annenberg and
Ford Foundations, launched the *Teachers for a New Era* (TNE)
Initiative, with the goals of catalyzing reform in teacher
education, consolidating knowledge about excellence in teacher
preparation, and strengthening public confidence in university
teacher education programs (UTEPs). In 2002, four universities,
including California State University, Northridge (CSUN), were
invited to join TNE and awarded $5 million each over five years;
another seven joined in 2003. Our charges were:

1. To make teacher education a university-wide effort, specifically engaging arts and sciences faculty in improving subject-matter learning for potential teachers;
2. To strengthen the clinical-practice experience, by intensifying partnerships and the theoretical alignment between the university and K–12 schools;
3. To ground continual program improvement in evidence of the impact of programs on the teaching practices of graduates and the learning of their pupils.

This last charge also addresses the goal of public confidence: If UTEPs can establish systematic processes for gathering evidence of their impact on teachers and pupils, they become, for the first time, positioned to demonstrate their value to the enterprise of teacher development.

TNE efforts at CSUN have involved over 200 people through multiple projects addressing the charges. These include faculty members and administrators at CSUN and area community colleges, local K–12 school personnel, preservice teachers, and area business people. This article describes the work of one committee, tasked with the third charge, to systematize the collection of evidence of programmatic impact. In the 2006–2007 school year, we conducted an evaluation study of the secondary-level mathematics-credential program. The main purpose of this article is not to convey our particular results, which may be of limited relevance to other programs, but to explain the study's design and rationale and what we learned, to inform other programs interested in conducting evidence-based self-evaluations.

The Challenges of Evaluating a Program by the Performance of Its Graduates

A teacher educator need only consider for a few minutes the prospect of evaluating a credential program on the basis of its graduates' teaching practices before problems become apparent. The few listed here weighed heavily on our team and shaped our study's design.

1) *Scarce resources.* As a public, regional, masters university, time and money for research is limited and we have no doctoral students to assist. TNE provided initial funding, but we hoped to design an ongoing evaluation that could be sustained with normal university resources.

2) *What to look for.* It was obvious that we would need to observe graduates "in the field," teaching in classrooms. It was not obvious what to look for that would convince not only department members but also the broader public of our program's impact.

3) *How to discern program impact.* A host of factors shape what teachers do in the classroom. Some arise on the job, including curricular mandates, professional development and mentoring, characteristics of pupils and their families, and available resources. These can support or compete with factors teachers bring to the job: personal philosophies and school experiences; knowledge, skills, and dispositions gained in UTEPs; and personal or family situations. How would we know whether to attribute certain teaching practices to our credential program?

These challenges make clear why few UTEPs have engaged in this sort of evaluation. In our study, we reached partial resolution to these problems but cannot claim to have solved them.

Study Design

The core study team comprised two faculty members in the Department of Anthropology and three in the College of Education. Only the first author was in Secondary Education, the department that housed the program being evaluated. We also employed two research assistants. The core team designed the study to serve two main purposes: We aimed to provide specific evidence for the Department of Secondary Education about its graduates' teaching, to inform program improvement; and to demonstrate that an "in-house" evaluation of a credential program, based on evidence from the field, was possible, and to present one method.

Our research questions were:

- To what degree do new credential-program graduates implement program-emphasized teaching practices in their classrooms?
- What factors facilitate or impede the implementation of program-emphasized practices?
- Is the program improving over the years in terms of this implementation?

Participants

An early and crucial decision was how many graduates to include in the study and how often to observe them. With limited resources, this was a breadth-versus-depth decision. We opted for breadth, observing as many participants as we could, once or twice each, because we were interested in the impact on *all* graduates, and because we were anxious to convey an evaluation of the program, not of individual teachers. While we understood that no single teacher's practices would be fairly captured in one or two observations, we hoped the large number of "snapshots" would paint a reasonably accurate picture of our graduates in aggregate.

Another key decision was which graduates. New teachers can be overwhelmed by the challenges of classroom and time management and becoming familiar with school policies and facilities. These challenges may prevent them from utilizing practices they have learned in a UTEP, would like to use, and may use in later years (Richardson, 1996). Observing teachers with more experience might give a truer picture of how they "really" teach, but it would be harder to credit our program for what we observed, because of intervening professional development experiences in the years since graduation. Our ultimate decision to observe first- and second-year teachers had two advantages: It is easier to track down more recent graduates, and they are likelier to still be teaching. We invited all 31 first- and second-year graduates of our traditional single-subject mathematics-credential program to participate. Of the 27 we were able to contact, 21 were fulltime mathematics teachers. Twelve agreed to participate, but we eliminated two because of distance. Thus, ten teachers were observed and interviewed at least once; six were observed twice, for a total of 16 observations/interviews. We secured permission from each participating teacher's principal; thus, it was important to disguise vignettes or quotes in reported results. Participants were paid $50 for each observation/interview.

Observations and Interview

We made two kinds of observations in the classroom. We recorded instances of the teacher's use of specific practices that the program emphasized; how these were designated is discussed later. Because these designated practices were exemplary and somewhat advanced, we did not expect to see them used frequently, and by documenting only them we would risk missing most of what the teachers did during the lessons. Therefore, we decided also to keep a running record of the teaching/learning modes, noting the clock time of each mode change. We developed a manageable list of modes to capture critical classroom-environment qualities known to impact learning: individual versus collaborative work, teacher versus pupil talk, and teacher presentation versus pupil discovery or idea generation (e.g., Boaler, 1997; Hiebert et al., 1997). The final list of modes was:

1) Teacher presentation of behavioral directives or task instructions
2) Teacher presentation of math content (minimal or no pupil input)
3) Pupil demonstration (more significant than verbal answer from seat)
4) Whole-class discussion to review/apply learned procedures (significant pupil input)
5) Whole-class discussion to co-construct new concepts or procedures
6) Pair or small-group work practice of learned procedures

7) Pair or small-group concept development (e.g., discovery activity/lab/project)
8) Individual practice of learned procedures
9) Individual concept development (e.g., discovery activity/lab/ project)
10) Activity or discussion unrelated to math or implementation of the lesson.

Designating the specific practices to observe for was more problematic. One option was to derive a list from the research literature about effective mathematics-teaching practices. But this might have clouded our view of our program's impact, because not all of those practices would necessarily have been emphasized in our program, due to limited program duration and the beginning level of our students. Similarly, existing observation protocols, such as the Reformed Teaching Observation Protocol (Arizona Collaborative for Excellence in the Preparation of Teachers, 2000) and the National Survey of Science and Mathematics Education Mathematics Questionnaire (Horizon Research, Inc., 2000), were far more extensive than our resources would support and assessed more practices than appropriate for new graduates. Instead, we convened a group of faculty members who taught in the math-credential program to generate a list of practices they believed were emphasized in courses and fieldwork. They focused on research-based practices that seemed reasonable to expect from a new teacher and could validly be observed in one lesson; hence, they excluded assessment practices. From this list, the core team developed an observation protocol, combining like practices, eliminating less significant ones, to attain a manageable number. Published protocols were referenced; indeed, all items on our protocol have close counterparts in other protocols. But for the sake of program evaluation and faculty buy-in, it was important that our decision to include each practice was based on faculty agreement that it was emphasized in our program.

Table 1 lists the practices as described in the observation protocol and gives example citations for scholarship linking the practice to enhanced pupil learning. During the lesson, the observer(s) described in field notes any instances of these designated practices. Afterwards, the observer(s) rated each practice as being a significant, a marginal, or no part of the overall lesson. These ratings were holistic, with the marginal rating used for lessons where instances of the practice were very short in total duration (e.g., one practice problem permitting a choice of method) and/or superficial (e.g., a teacher's side remark about a real-world connection).

Table 1

Practices as Listed on the Observation Protocol, with Examples of Supporting Scholarship

Practice	Example of supporting scholarship
1) Teacher asks question or poses task with a **high level** of cognitive demand	Kilpatrick, Swafford, and Findell (2001)
2) Pupil given **authority to judge** the mathematical soundness of publicly presented solution or method	Hiebert et al. (1997)
3) Teacher **connects** (or poses task that prompts pupil to connect) the featured math topic to **another math** topic	Hiebert and Carpenter (1992)
4) Teacher **connects** (or poses task that prompts pupil to connect) the featured math topic to **another academic** topic	Lehrer, Schauble, Strom, and Pligge (2001)
5) Teacher **connects** (or poses task that prompts pupil to connect) the featured math topic to a **real-life** situation or object	Boaler (1997)
6) Pupils allowed or encouraged to **choose** among solving **methods** or present alternative methods	Bransford, Brown, and Cocking (1999)

Practice	Example of supporting scholarship
7) Teacher or pupils use **technology, manipulatives**, body movement, or other nonverbal support for a math concept	Kilpatrick et al. (2001)
8) Specific attention paid to **language**, i.e., writing, reading, or speaking skills	Borasi and Siegel (2000)
9) Teacher uses or encourages pupil to use multiple **forms of representation** for the same problem	Bransford et al. (1999)

The designated practices are not basic teaching methods but reform-oriented, nontraditional strategies that are difficult to carry out effectively (Tarr, Reys, Reys, & Chavez, 2008) and probably require more intensive training than can be accomplished in a one year post-baccalaureate credential program. Therefore, we decided not to try to assess the quality of their implementation or their impact on pupil learning, even though the latter was part of the third TNE charge. We felt any level of implementation of these practices by new teachers would be notable, but that it would be unreasonable to expect that implementation to result in observable pupil outcomes.

Observations were scheduled with the teacher. We requested a first-year algebra class or the closest course the teacher taught, to attain some commensurability across participants. Also, Algebra I is of great concern in California: required for a high school diploma and with a high failure rate, it has become the "bottleneck" course. We asked only for a typical lesson—nothing special—and that there not be a test that period. Surprise visits were not an option; the schools would not have permitted them, and we wanted to offer participants some agency over the process. We were unconcerned about teachers "putting on a show." Indeed, if a teacher were to perform atypically, it would not entirely represent a "false positive." If a recent graduate could anticipate what practices CSUN researchers would hope to see (we did not tell participants we were targeting specific practices), understood how these differed from her usual methods, and were able to incorporate them into a lesson, we could justifiably interpret this as a form of programmatic impact. It would at least indicate that: a) the teacher had learned what CSUN considered good teaching practices, b) she was able to implement them, c) the stretch to incorporate them into her normal classroom environment was not so far as to be impossible, and d) she was disposed to trying new practices. On the other hand, with the literature on teacher change showing strong resistance even among veteran teachers to adopt new practices, despite significant professional development and explicit encouragement from administrators (Clarke, 1994), it seems unlikely that a new teacher would be willing or able to drastically alter her performance for a researcher who has not explicitly specified a "good performance."

Prior to the observation, each teacher filled out a form that asked for background information (education, experience, and languages spoken) and about the lesson and class to be observed (demographics, lesson objectives, and any special features or conditions). After the observation, on the same day, we conducted a semi-structured interview with each participant to try to understand why we did or did not see the designated practices implemented during the lesson. Grossman, Smagorinsky, and Valencia (1999), in their framework for studying teaching, laid out multiple variables to consider in explaining new teachers' practice appropriation—more than a single study could attend to, especially a necessarily resource-lean program evaluation. Yet we knew it was crucial to examine contextual variables and not presume that whether and how our graduates appropriated program-emphasized practices were solely functions of the quality of our instruction. In particular, we inquired about the impact of teaching settings (collegial and administrative support, demographics, professional development, material resources, etc.), personal background (motivations for teaching and prior experiences), and personal interpretations and affective impressions of our UTEP. Examining these from the teachers' perspective aligns with the stance of Grossman et al. that settings are personally construed and that a teacher's interpretation of these variables determines her motives and impacts practice.

Training

Training our team to use the observation protocol was accomplished with videotaped lessons, adapting calibration methods from large-scale classroom-observation studies such as the TIMSS Video Study (National Center for Education Statistics, 2003). We all viewed the same portion of a video and noted instances of the designated practices and teaching modes. Then we discussed our observations until we reached consensus about whether each instance indeed exemplified the practice or mode. We repeated this exercise several times, also using the first few sessions to refine the protocol and our list of examples of each practice and mode, until we were consistently reaching a high level of agreement (about 6 video segments). Calibration was further achieved by sending observers in pairs for their first two observations; afterwards, the pair reached consensus about modes and significance ratings. During the fall and spring that observations and interviews occurred, the team met periodically to discuss uncertain ratings, refine the observation protocol's example bank, and discuss emerging trends in our findings.

Data Analysis

Analysis of the quantitative data was relatively simple: We totaled the time spent in each teaching mode and calculated each as a percent of the total time, to produce an "average" lesson profile for our graduates. For each designated practice, we reported the percent of observed lessons that employed it at each significance level. The qualitative data were analyzed three ways. First, interview data were read for factors that supported or constrained teaching, some of which had been anticipated and built into the interview prompts, others of which emerged, following a grounded theory approach (Creswell, 1998). When multiple readers had agreed on a coding scheme of factors, the data were reread and coded, and numbers of teachers citing each factor were tallied. Second, case studies (Creswell, 1998) were developed for 6 of the teachers, selected to represent a range of use of the designated practices. Drawing from the interviews, observations, and information forms, these interpretive narratives were intended to provide holistic illustrations of how the various factors interacted to influence teaching. Finally, the 10 teachers were categorized by level of practice implementation, and all data were reanalyzed for differential patterns of influential factors across (relatively) low, medium, or high implementers.

Findings

In general, we found our graduates implementing the designated teaching practices to a low degree. The most frequently observed of these practices was giving pupils the authority to judge the mathematical soundness of a solution or method—a significant part of 25% of the lessons. Posing tasks with a high level of cognitive demand was significant in 19% of the lessons. These two practices played at least a marginal role in half of the lessons, as did the use of technology or manipulatives. All other designated practices were observed less often.

The modes employed in our graduates' "average" mathematics lesson reflected relatively traditional teaching. For nearly 1/3 of the duration of this average lesson, pupils practiced learned procedures ("seatwork"), sometimes being allowed but not explicitly encouraged to consult with peers. For nearly 1/4 of the lesson, teachers presented mathematical content, taking minimal or no pupil input. Teachers gave behavioral directives or task instructions for 12% of lesson time. All other modes were each used less than 10% of the time.

We had anticipated this picture of our graduates' first and second years of teaching. Other, larger-scale studies (e.g., Jacobs et al., 2006; Weiss, Pasley, Smith, Banilower, & Heck, 2003) have documented the low degree of implementation of reform practices in mathematics classroom by teachers at *all* levels of experience. New teachers may be particularly apt to fall back on traditional methods, with preservice teacher education constituting too weak an intervention to overcome teachers' experiences as students in traditional classrooms (Ball & Cohen, 1999) or the pressure to align with the culture of their initial job placement (Grossman et al., 1999). We view these as baseline data, with the real value of the study coming in its continuation over the years, as the results, ideally, feed into and reflect program improvement. Further, if specific program improvements are followed by observations of increased frequency of related practices, it will arguably be evidence that UTEPs impact teaching. Of special interest to us is the effect of a significant program change: Since 2006–07, we have instituted a capstone performance assessment that requires credential candidates to reflect extensively on a unit they plan and teach, including a videotaped lesson, pupil work samples, and their efforts to develop pupils' academic language.

Also important findings are the factors that participants cited to explain how they teach. We gained specific feedback about aspects of our program and how our graduates, now one or two years into teaching, saw their value. What participants perceived as supports (mainly collegial and administrative support) and constraints (mainly managerial tasks and limited time) at the school site echoed existing research (e.g., Gold, 1996; Darling-Hammond, 1990) about the experience of new teachers.

Repeating the Study

This study demonstrated that an evidence-based self-evaluation of a credential program is possible and can be carried out with minimal resources. In 2008–09, we are repeating the study with a few changes. First, because second observations/ interviews of the same teacher appeared to have little effect on the overall results, we will only visit each teacher once but double the pool of potential participants, now targeting graduates in their first through *fourth* years of teaching, thereby revisiting the 2006–07 participants. We have also modified the interviews to yield information more easily interpreted in terms of programmatic improvement. In 2006–07, teachers were asked about constraints and supports for their teaching in general, leaving us unsure whether time pressure, for example, specifically constrained the use of any designated practices or was just an overall impediment. Questions for 2008–09 more directly target factors that influence particularly the implementation of the practices. Other changes reflect the need to reduce the cost of the study to make it feasible on an ongoing basis. The observation protocol and calibration training procedures are unchanged, but only two researchers will conduct the study. Data analysis will also remain the same except that no further case studies will be developed. Our focus will continue to be on influential factors in the teachers' UTEP preparation and in their current teaching situation, to help the department understand how to more effectively promote good teaching practices and prepare new teachers to navigate constraints in the field.

References

Arizona Collaborative for Excellence in the Preparation of Teachers. (2000). Reformed Teaching Observation Protocol. Retrieved August 21, 2009, from http://cresmet.asu.edu/instruments/RTOP/index.shtml

Ball, D. L., & Cohen, D. K. (1999). Developing practice, developing practitioners: Toward a practice-based theory of professional education. In L. Darling-Hammond & G. Sykes (Eds.), *Teaching as the learning profession: Handbook of policy and practice* (pp. 3–32). San Francisco: Jossey-Bass.

Boaler, J. (1997). *Experiencing school mathematics: Teaching styles, sex, and setting.* Philadelphia: Open University Press.

Borasi, R., & Siegel, M. (2000). *Reading counts: Expanding the role of reading in the mathematics classroom.* New York: Teachers College Press.

Bransford, J. D., Brown, A. L., & Cocking, R. R. (Eds.) (1999). *How people learn: Brain, mind, experience, and school.* Washington, DC: National Academy Press.

Clarke, D. (1994). Ten key principles from research for the professional development of mathematics teachers. In D. B. Aichele & A. F. Coxford (Eds.), *Professional development for teachers of mathematics: 1994 Yearbook* (pp. 37–48). Reston, VA: National Council of Teachers of Mathematics.

Creswell, J. W. (1998). *Qualitative inquiry and research design: Choosing among five traditions.* Thousand Oaks, CA: Sage Publications.

Darling-Hammond, L. (1990). Teachers and teaching: Signs of a changing profession. In W. R. Houston (Ed.), *Handbook of research on teacher education* (pp. 267–290). New York: Macmillan.

Gold, Y. (1996). Beginning teacher support: Attrition, mentoring, and induction. In J. Sikula, T. J. Buttery & E. Guyton (Eds.), *Handbook of research on teacher education* (2nd ed.; pp. 548–594). New York: Simon & Schuster Macmillan.

Grossman, P. L., Smagorinski, P., & Valencia, S. W. (1999). Appropriating tools for teaching English: A theoretical framework for research on learning to teach. *American Journal of Education, 108*, 1–29.

Hiebert, J., & Carpenter, T. P. (1992) Learning and teaching with understanding. In D. A. Grouws (Ed.), *Handbook of research on mathematics teaching and learning* (pp. 65–97). New York: Macmillan.

Hiebert, J., Carpenter, T. P., Fennema, E., Fuson, K. C., Wearne, D., Murray, H., Olivier, A., & Human, P. (1997). *Making sense: Teaching and learning mathematics with understanding.* Portsmouth, NH: Heinemann.

Horizon Research, Inc. (2000). 2000 National Survey of Science and Mathematics Education Mathematics Questionnaire. Retrieved December 30, 2008, from http://2000survey.horizon-research.com.

Jacobs, J., Hiebert, J., Givvin, K., Hollingsworth, H., Garnier, H., & Wearne, D. (2006). Does eighth-grade mathematics teaching in the United States align with the NCTM Standards? Results from the TIMSS 1999 video studies. *Journal for Research in Mathematics Education, 37*(1), 5–32.

Kilpatrick, J., Swafford, J., & Findell. B. (2001). *Adding it up: Helping children learn mathematics.* Washington, DC: National Academy Press.

Lehrer, R., Schauble, L., Strom, D, & Pligge, M. (2001). Similarity of form and substance: From inscriptions to models. In D. Klahr & S. Carver (Eds.), *Cognition and instruction: 25 years of progress* (pp. 39–74). Mahwah, NJ: Lawrence Erlbaum Associates.

National Center for Education Statistics. (2003). *Teaching mathematics in seven countries: Results from the TIMSS 1999 Video Study,* (NCES 2003–013 Revised). Washington, DC: U.S. Department of Education.

Richardson, V. (1996). The role of attitudes and beliefs in learning to teach. In J. Sikula, T. J. Buttery & E. Guyton (Eds.), *Handbook of research on teacher education* (2nd ed.; pp. 102–119). New York: Simon & Schuster Macmillan.

Tarr, J. E., Reys, R. E., Reys, B. J., & Chavez, O. (2008). The impact of middle-grades mathematics curricula and the classroom learning environment of student achievement. *Journal for Research in Mathematics Education, 39*(3), 247–280.

Weiss, I. R., Pasley, J. D., Smith, S. P., Banilower, E. R., & Heck, D. J. (2003). *Looking inside the classroom: A study of K–12 mathematics and science education in the United States.* Chapel Hill, NC: Horizon Research, Inc.

Julie Gainsburg, Ph.D., is Associate Professor of Secondary Education at California State University, Northridge. She earned her Ph. D. in Curriculum and Teacher Education at Stanford University in 2003. Her interests are mathematics-teacher education and development, and the relationship between school mathematics and everyday and workplace mathematics.

Carrie Rothstein-Fisch, Ph.D., is Professor of Educational Psychology and Counseling at California State University, Northridge. She earned her Ph.D. in Educational Psychology, with an emphasis in Early Childhood and Developmental

Studies, at the University of California-Los Angeles, in 1983. Her specialties are cultural diversity, early childhood education, and teachers' professional development.

Suzanne Scheld, Ph.D., is Assistant Professor of Anthropology at California State University, Northridge. She earned her Ph.D. in Cultural Anthropology at The City University of New York, The Graduate Center in 2003. Her interests are urban education, youth culture, and the politics of public space in the U.S. and West Africa.

Michael Spagna, Ph.D., is Dean of Michael D. Eisner College of Education at California State University, Northridge. He earned his Ph.D. in Special Education at the University of California at Berkeley and San Francisco State University in 1991. His interests are: (a) the design and implementation of cognitive, academic, social and emotional methodologies to be used with all students, particularly struggling learners; (b) the design, implementation, and evaluation of a variety of teacher preparation models; and, (c) the integration of interprofessional and transdisciplinary service activities into college-wide preparation programs and curriculum.

Christina von Mayrhauser, Ph.D., formerly Associate Professor of Anthropology at California State University, Northridge, is currently serving as Associate Dean of that institution's College of Social and Behavioral Sciences. She received her Ph.D. in Anthropology from University of California-Los Angeles in 2000. Current interests focus on anthropological applications in higher education research.

Rhodes, G. and Wilson, P. S.
AMTE Monograph 6
Scholarly Practices and Inquiry in the Preparation of Mathematics Teachers
© 2009, pp. 121–135

8

Field Experiences: An Opportunity for Mentor Learning

Ginger Rhodes
University of North Carolina Wilmington

Patricia S. Wilson
University of Georgia

Mentoring student teachers provides rich opportunities for in-service teachers to learn about their teaching practices. In this paper we share our efforts to design professional learning experiences for in-service teachers within the context of student teaching. We identify three characteristics of professional learning that were critical in creating opportunities for mentor learning: learning within the context of practice, learning from students' mathematical thinking, and learning from collaboration. We also share challenges to consider when orchestrating professional development experiences within field experiences for in-service teachers.

Learning a profession requires formal study and practical experiences, and teaching is no exception. Field experiences have long been an integral part of teacher preparation programs and students eagerly embrace the opportunity to work in schools and learn from veteran teachers. The expectation is that the mentor teacher will dedicate a substantial amount of time and effort, and the student teacher will be the learner. We want to challenge the notion that mentoring is primarily for the benefit of student teachers. Our claim is that carefully planned field

experiences can provide an opportunity for substantial learning
for teachers who serve as mentors as well. We argue that
reprioritizing the responsibilities of mentors can increase their
own learning. Rather than envisioning mentoring as providing
advice and supervision, we see mentoring as establishing a
collaborative working relationship among colleagues (Awaya, et
al., 2003) so that both mentors and protégés become more
reflective and learn more about teaching mathematics. Creating
such an opportunity is not easy because traditional roles within
field experiences are firmly established, and learning goals
within field experience are traditionally expressed in terms of the
protégé.

Mentoring a student teacher is difficult to do well (Sudinza,
Giebelhaus, & Coolican, 1997) and often the responsibilities are
not well-defined (Carver & Katz, 2004). Some mentor teachers
seek to model good teaching and others prefer to guide the
practice of their protégé by posing challenging questions
(Wilson, Anderson, Leatham, Lovin, & Sanchez 1999), but few
approach their mentoring work as a learning opportunity
intended for themselves. Although teachers claim to learn how to
teach from experimenting with their practice and talking with
colleagues (Wilson, Cooney, & Stinson, 2005), the structure of
most mentor programs within field experiences is not built on
collaboration. By definition, the mentor has experiences that the
protégé lacks. We believe that field experiences can be
structured so that the mentor's experiences enhance collaborative
learning and therefore provide a professional growth opportunity
for both the mentor and protégé.

In university teacher preparation programs, field
experiences, including student teaching, are typically designed to
provide learning experiences for the prospective teachers.
Mentor teachers may develop knowledge of their own practice
through hosting student teachers (Jaworski & Watson, 1994), but
this is usually a consequence of the experience rather than by
design. While some programs have successfully provided
training for mentors (Giebelhaus & Bowman, 2000), the
underlying philosophy of these programs is to support the
development of mentor teachers' supervisory skills in order to

improve their work with student teachers. These skills include giving advice, observing and providing feedback, and initiating reflective discussions with student teachers. Mentor teacher study groups have also been shown to help teachers develop their mentoring practice as well as promote inquiry-oriented learning (Carroll, 2005). We are extending ideas about mentoring by suggesting that programs need to support and encourage mentor teachers to work collaboratively with student teachers and colleagues to develop their own *teaching* practice.

In the following sections we share what we have learned from our efforts to improve field experiences in our teacher education program for mathematics teachers. We purposely designed experiences that provided learning opportunities for mentor teachers and university teachers in addition to student teachers. We planned experiences in which mentor teachers discussed their practice with colleagues as well as student teachers, and we encouraged collaborative work. Teachers who collaborate with their peers are likely to make and sustain changes in their instructional practices (Little, 1982), so we wanted to establish an atmosphere of collaborative learning that would support mentors' work with student teachers and would continue after the student teachers depart.

Consistent with research on successful professional development (Loucks-Horsley, Love, Stiles, Mundry, & Hewson, 2003; Smith, 2001), we identified three characteristics of professional learning that we think are important to consider in designing field experiences that promote mentor learning. We offer examples from our work that highlight these characteristics, and identify challenges that exist when implementing these modifications during field experiences. We note the importance of a dispositional shift that contributes to making field experiences valuable learning experiences for mentors.

Professional Growth within Field Experiences

Partnerships in Reform in Mathematics Education (PRIME) is a multi-level professional development project for three

groups of teachers: 1) student teachers preparing to be high
school mathematics teachers, 2) practicing teachers who serve as
mentors, and 3) university teachers who organize the
professional development experiences. PRIME has been part of
the program for preparing secondary mathematics teachers at the
University of Georgia since 1998. We began with six schools, 18
student teachers, 18 mentor teachers, and five university teachers
and grew to 13 schools in 2006, when we had 51 student
teachers, 54 mentors, and 14 university teachers. The project was
founded on the idea that the context of student teaching can
provide meaningful learning opportunities not only for student
teachers, but also for university teachers and mentor teachers.

Student teachers were placed in groups of two to five at each
participating school, but the structure varied by school. In many
cases one student teacher was paired with one mentor teacher,
while in other cases there were two mentor teachers with one
student teacher or two student teachers with one mentor. The
community of learners, which we denoted as a *Cluster,* included
the mentor teachers, student teachers, and one university teacher
at each school site. The Cluster met for 30 to 45 minutes once a
week during the student teaching semester to discuss and
examine *the work of mathematics teaching.* We used this phrase
to emphasize our focus on mathematics. The primary purpose of
the weekly Cluster meeting at each school was to provide a rare
opportunity for mentor teachers to discuss, analyze, and reflect
on their practice with a group of colleagues. We encouraged
members of the Cluster to move beyond the traditional student-
mentor teacher dyad and work as a collaborative group to
enhance their knowledge of mathematics for teaching.

The role of the university teacher was to serve as a catalyst
for collaborative group work. Since the Cluster meeting time was
short, the university teacher organized and structured the given
time to encourage the group to focus on issues related
specifically to teaching mathematics, rather than general issues
of teaching. Each Cluster evolved over the semester and was
unique based on the university teacher's involvement and the
dynamics of the group. The Cluster discussions were most
valuable when the discussion centered on events from the

teachers' mathematics classrooms and specifically on students' mathematical thinking. Some examples of activities included: examining student work, discussing a student's mathematical question, and solving a mathematics task. In one successful Cluster, the university teacher planned activities for the group during the first few meetings, and then encouraged other teachers to become involved in the planning process during later meetings. For example, during a Cluster meeting early in the semester, the university teacher brought samples of student work and led a group discussion about the work. Later in the semester, the university teacher worked with one student-mentor teacher dyad to collect student work samples from their classroom and to develop focus questions for the meeting. During the Cluster meeting the student-mentor teacher dyad took a leadership role by posing their questions and leading the analysis of the student work.

We organized a large meeting, called the Gathering, during the fall semester in which all of the mentor teachers and university teachers came together to think about possible activities for working collaboratively with student teachers and for the Cluster meetings. At the Gathering, teachers were asked to analyze samples of high school student work and then discuss the students' mathematical understanding that could be inferred from the samples. There were three additional Gatherings during the student teaching semester that included the student teachers. The purpose of these meetings was to engage teachers in activities that could be emulated in their Cluster meetings and to provide opportunities for the teachers to discuss their teaching and mentoring practices with colleagues from other schools.

An Opportunity that Promoted Mentor Learning

Based on PRIME work, we offer an example of a situation that occurred during a Cluster meeting that provided a learning opportunity for mentor teachers and then discuss three characteristics of the situation that professional development researchers have identified as contributing to the learning process. During a Cluster meeting at one high school, a group of three mentor teachers, three student teachers, and a university teacher watched a ten-minute videotaped lesson on synthetic division from a student teacher's class. The university teacher encouraged this activity by asking the student teacher to prepare and share a short video of her classroom practice. Before sharing the video the student teacher described her concern that she had not let the students become involved in thinking about the mathematics, but instead she had led them through a step-by-step process of synthetic division without allowing them to develop or question the concept. Her stated concern focused the group's observation, creating a common purpose for viewing the video.

After watching the video clip the teachers immediately began discussing what they had observed. The experienced teachers were able to recognize subtle student questions that had not seemed significant to the student teacher at the time of the lesson. The group was attracted to one student's question in the video that indicated that the high school student was looking at a relationship between two numbers in the algorithm. One mentor noted that this was an insightful comment because it was very difficult for students to make such connections. As a group, the teachers focused on this high school student's mathematical thinking and recognized that the student's observation provided an opportunity to teach mathematical connections even when the lesson was about a procedural task. The discussion progressed as they explored a student teacher's question, "Why do we teach synthetic division?" Various responses to this question led to a discussion of the use of synthetic division in factoring polynomials of degrees greater than two, and then jumped to the relationship between synthetic division and long division. The conversation transitioned to connections among mathematical topics such as the derivative of a function and its critical points. At this point the discussion had shifted from the mathematics of high school students to the mathematics of the teachers, and they began to see implications for the curriculum.

This example of a Cluster meeting discussion illustrated the excitement that can be generated by using actual episodes from practice and by focusing on the mathematical thinking of high school students. The discussion was nonlinear due to the spontaneous interest and participation of teachers who taught a variety of classes. A discussion that began with a procedural task developed into a collection of mathematical insights. Each teacher was exploring mathematics, seeing new ideas for teaching mathematics, and contributing to the conversation. This kind of collaboration promoted mentor learning in addition to student teacher learning. Such mentor interaction has the potential to lead to changes in practice and thinking. One of the mentor teachers captured the impact of such learning experiences:

> Sometimes I just need a jumpstart to be more reflective about my teaching and to take more risks. It's very easy to get into a comfortable mode of doing the same things each year. This year has challenged me to re-think what I am teaching, why I am teaching it and if there are better ways to get my students to learn the things I want them to learn. (Written Reflection, May 2005)

Characteristics of Opportunities that Promote Mentor Learning

Our Cluster example exhibited the three key characteristics consistent with recommendations for high-quality professional development (Loucks-Horsley et al., 2003; Smith, 2001): structured within the context of practice, focus on students' mathematical thinking, and collaboration within the group. These characteristics were not only prevalent in successful Cluster meetings but they also suggest ways that traditional student teaching activities could be modified or adapted to promote mentor learning.

By structuring events within the context of practice we were assured that the experience would reflect the excitement, the constraints, and the reality of teaching mathematics. In the previous example, both mentor teachers and student teachers were enthusiastic about the discussion of synthetic division because it was related to their practice as mathematics teachers. The teachers discussed a student's mathematical thinking and eventually their own mathematical thinking connected to synthetic division. They made strides in the difficult task of switching the focus of their conversation from "What should I do as a teacher?" to "What are my students thinking and how can I build on their thinking?" Notably, the example from the Cluster meeting began with a concern from one student teacher, which quickly became the concern and focus of the entire group. The diversity of the group of teachers then encouraged a variety of ideas to surface about that concern. As a result, the mentor teachers weren't focused solely on "teaching" the student teachers; they were examining teaching and learning mathematics for their own learning with colleagues. We have found that such collaboration is a key characteristic that distinguishes a mentor who is learning from a mentor who is advising.

Challenges in Facilitating Mentor Learning

Our experience with the PRIME project illustrates the potential learning opportunities for mentor learning, as well as the difficulties that may arise when trying to achieve such a goal. Ensuring that mentoring work is set in the context of practice, focuses on students' mathematical thinking, and promotes collaboration among fellow teachers may present unique challenges for the teachers involved.

First and foremost, both the mentor and student teacher must approach their joint work as learners. One university teacher commented, "I'm becoming more convinced that the mentor teachers have to take ownership for their professional development. And just because they're in the PRIME program, if they don't really want to improve themselves it's not going to happen" (Interview, November 2005). There are failures in the student-mentor dyad relationship when "the mentor is unable or unwilling to work collaboratively with a particular student, or the student is unable to engage the mentor as an equal" (Awaya et al., 2003, p. 55). The teachers must be willing to investigate their practice, contribute and share ideas, and consider each other's ideas. This approach to mentoring is difficult for some teachers because the beginning teacher is hungry for advice and the mentor teacher has an abundance of experiential knowledge and wants to be helpful. For example, one PRIME mentor teacher thought the project "should address problems and situations that arise during their [student teachers'] student teaching time. Mentor teachers and PRIME leaders can then offer advice on how to correct the problem situation" (Written Reflection, May 2005). This disposition is typical in situations in which the mentor is charged with the responsibility of providing advice. Yet Feiman-Nemser (2001) described a different view, referring to one mentor teacher's desire to model his wondering about teaching as a way to improve his own teaching, as well as a mentoring strategy. This mentor teacher would share a question that he had about his teaching practice with the novice teacher, and then together they would examine data to think about the question. We do not intend to suggest that mentor teacher advice is not valuable or appropriate for the student teachers' professional growth. We are stating that mentoring activities that focus solely on giving the novice teacher advice limits the collaborative learning opportunities for mentor teachers. Furthermore, when the mentor teacher models wondering as a teaching practice, the novice teacher learns to investigate and improve her practice by doing the same.

Second, limited time is a constraint that many teachers deal with daily. We wanted to encourage PRIME teachers to jointly analyze artifacts of their practice and to collaboratively discuss students' mathematical thinking. In order for such activities to be successful the teachers must carve out time to dedicate to their joint work. The university teacher at one school site commented, "Friday was the only day that the teachers could meet... but it got to the point where they just fizzled out" (Interview, November 2005). At this particular school site the teachers found it difficult to meet after school, and consequently the time allotted for collaborative work was limited. Some Clusters tried to meet before school or during lunch. Yet the Clusters that met during lunch found it difficult to stay focused on collaborative work due to multiple interruptions and teachers eating. Time constraints are genuine problems that need creative solutions.

Finally, to orchestrate experiences such as the ones we have described, there needs to be strong leadership. There needs to be a leader who coordinates activities among the teachers and supports the teachers in their work. The leadership in the project PRIME was provided by the university teachers who organized, structured, and facilitated the Cluster meetings. Both the mentor teachers and student teachers contributed to the planning and direction of the Cluster meetings, but university teachers set up the meetings and encouraged teachers to bring teaching artifacts from their classrooms to the meetings. But simply organizing the meeting is not sufficient for successful collaborative work. The leader should promote the three characteristics of professional learning as well as be an active member of the collaborative group. In one Cluster, the university teacher collaborated with the student-mentor teacher dyads to plan and teach a lesson. Following the activity, the group of teachers reflected on the experience. In this example, the university teacher approached her work with the teachers as a learner.

These three challenges need careful attention when organizing and planning mentoring activities. As teacher educators we need to consider our roles within field experiences and ways that we can work with and support mentor teachers as they learn with novices about their teaching practices.

Improving Opportunities for Mentor Learning

While much of the collaborative work that PRIME facilitated occurred within Cluster meetings and the Gatherings, PRIME also encouraged that work within the student-mentor teacher dyads. In the following section, we share a common mentoring activity from the PRIME project that occurred between mentor teachers and student teachers. While we believe these activities are valuable experiences for the student teacher, we claim that a purposeful modification of the activity can increase the extent to which the experience incorporates the three characteristics and consequently provides a professional growth opportunity for the mentor and student teacher. We also share an example from the project that reflects these modifications and discuss how this example exemplifies the three characteristics that we contend promote mentor learning.

Lesson planning is an essential part of the work of teaching. Creating lesson plans requires teachers to consider learning goals, potential learning activities, and the various ways that students might engage in and learn from those activities. In addition, teachers make lesson plans based on their understanding of mathematics, pedagogy, and students. We have found that joint lesson planning is a productive activity for student teachers and mentor teachers.

Many of the student-mentor teacher dyads discussed lesson plans on a regular basis. These discussions varied to the extent that each met the three characteristics of professional development and, ultimately, the extent to which the experience was a learning opportunity for the mentor teacher. The following discussion between a student teacher and mentor teacher was typical. The discussion began when a student teacher posed a question to his mentor teacher about how to teach factoring trinomials. The mentor teacher mentioned that factoring is difficult for students and that she had tried several approaches over the years. She proceeded to share three ways to factor trinomials. In her discussion the mentor teacher explained each method and noted the advantages and disadvantages of each method. The mentor teacher explained to the student teacher which method she thought he should use when he teaches factoring. Although the mentor teacher's advice may be helpful to the student teacher, the situation could be altered slightly to promote mentor teacher learning and improve the student teacher's ability to generalize to other topics. The following situation reflects a successful modification to the common lesson planning activity between mentor teachers and student teachers.

A mentor teacher wanted to use an unfamiliar mathematics task that incorporated the computer program Geometer's Sketchpad (GSP) (Jackiw, 1991) in class. The teachers wanted the students to construct the orthocenter, in-center, and circumcenter of a triangle using GSP. The mentor teacher and student teacher did the task together before they created their lesson plan. While completing the task both teachers commented on aspects of the activity that were difficult for them and thought about aspects that might be difficult for students. The teachers made reference to what students already knew about the underlying mathematics of the task. After the teachers completed the activity they discussed how to appropriately introduce the activity. The teachers continued the discussion by planning details of the lesson.

This opportunity for learning was set in the context of practice since the teachers were planning a lesson that they would use in their class. In addition, the teachers were considering the ways that their actions as teachers might affect student learning during the task. They were considering the students' mathematical thinking. The teachers completed the activity themselves and made connections between the students' starting points, potential student solution paths, and possible student struggles. In this example the mentor and student teacher were both taking an active role in planning a lesson; both teachers shared their ideas, listened and critiqued each other's ideas, and created the lesson jointly. The mentor teacher noted in her reflection:

> Having a student teacher to plan with forced me to explain my thoughts behind lessons, assessment activities, and projects or activities. It also gave me an opportunity to bounce ideas off someone who was receptive to new ideas and had ideas to contribute. [The student teacher] and I tried several different investigations in GSP that went very well. (Written Reflection, May 2005)

Concluding Remarks

"One intangible benefit for mentors is the opportunity to engage in conversations in which the mentor can learn from the new teacher" (Peterson & Williams, 1998, p. 732). Mentoring can provide a unique opportunity for teachers to collaborate and discuss the work of mathematics teaching. Through situations from PRIME we have shown ways that mentoring activities can provide worthwhile learning opportunities for both the novice and the mentor teacher. One PRIME teacher commented:

> It made me re-evaluate my approach to students' learning and my teaching styles. PRIME is a positive experience for not only student teachers but also for veteran teachers. Both the mentor teacher and the student teacher benefited from working with each other. Mentor teachers are forced to think more about the math. This proves to be advantageous for everyone involved. (Written Reflection, May 2005)

As the mathematics education community explores new ways to provide professional development for in-service teachers, we need to consider learning opportunities that exist within current structures. PRIME is an example in which hosting a student teacher provided a context for mentor teachers to investigate their own teaching practices. We think that what we have learned can inform a variety of mentoring programs, as well as research related to professional growth in teaching mathematics. In this paper we have presented further evidence that professional development experiences that are situated in practice, focused on students' mathematical thinking, and collaborative are difficult to achieve, but worth the effort.

References

Awaya, A., McEwan, H., Heyler, D., Linsky, S., Lum, D., & Wakukawa, P. (2003). Mentoring as a journey. *Teaching and Teacher Education, 19*, 45–56.

Carroll, D. (2005). Learning through interactive talk: A school-based mentor teacher study group as a context for professional learning. *Teaching and Teacher Education, 21*, 457–473.

Carver, C. L., & Katz, D. S. (2004). Teaching at the boundary of acceptable practice: What is a new teacher mentor to do? *Journal of Teacher Education, 55*, 449–462.

Feiman-Nemser, S. (2001). Helping novices learn to teach: Lessons from an exemplary support teacher. *Journal of Teacher Education, 52*, 17–30.

Giebelhaus, C. R., & Bowman, C. (2000, February). *Teaching mentors: Is it worth the effort?* Paper presented at the annual meeting of the Association of Teacher Educators, Orlando, FL.

Jackiw, N. (1991). *Geometer's Sketchpad.* Berkeley, CA: Key Curriculum Press.

Jaworski, B., & Watson, A. (1994). Mentoring, co-mentoring, and the inner mentor. In B. Jaworski & A. Watson (Eds.), *Mentoring in mathematics teaching* (pp. 124–138). London: The Falmer Press.

Little, J. W. (1982). Norms of collegiality and experimentation: Workplace conditions of school success. *American Education Research Journal, 19*, 325–340.

Loucks-Horsley, S., Love, N., Stiles, K. E., Mundry, S., & Hewson, P. W. (2003). *Designing professional development for teachers of science and mathematics* (2nd ed.). Thousand Oaks, CA: Corwin Press.

Peterson, B. E., & Williams, S. R. (1998). Mentoring beginning teachers. *Mathematics Teacher, 91*, 730–734.

Smith, M. (2001). *Practice-based professional development for teachers of mathematics.* Reston, VA: National Council of Teachers of Mathematics.

Sudinza, M., Giebelhaus, C., & Coolican, M. (1997, Winter). Mentor or tormentor: The role of the cooperating teacher in student teacher success or failure. *Action in Teacher Education, 18*, 23–25.

Wilson, P. S., Anderson, D. L., Leatham, K. R., Lovin, L. H., & Sanchez, W. B. (1999). Giving voice to mentor teachers. In F. Hitt & M. Santos (Eds.), *Proceedings of the twenty-first annual meeting of the North American Chapter of the International Group for the Psychology of Mathematics Education* (Vol. 2, pp. 811–817). Cuernavaca, Mexico: Universidad Autonoma del Estado de Morelos.

Wilson, P. S., Cooney, T., & Stinson, D. (2005). What constitutes good mathematics teaching and how it develops:

Nine high school teachers' perspectives. *Journal of Mathematics Teacher Education, 8*, 83–111.

Ginger Rhodes is an Assistant Professor in the Mathematics and Statistics Department at the University of North Carolina Wilmington where she teaches courses for pre-service and in-service mathematics teachers. Her research interests include teachers' instructional practices and their understanding of students' mathematical thinking.

Patricia S. Wilson is a Professor of Mathematics Education at the University of Georgia and a principal investigator of the NSF Center for Proficiency in Teaching Mathematics. Her research interests are focused on mathematics teacher development, including the mathematical knowledge that is useful for teaching mathematics at the secondary level.

Elliott, R.
AMTE Monograph 6
Scholarly Practices and Inquiry in the Preparation of Mathematics Teachers
© 2009, pp. 137–152

9

Preservice Mathematics Teachers Learning to Inquire Into Their Practice: Cases of Trying On Reform

Rebekah Elliott
Oregon State University

This chapter examines secondary mathematics preservice teachers' (PSTs) written cases of their classroom practice. This assignment was a means of developing PSTs' skills and capacity to inquire into their practice. Writing a case as part of a methods class, the PSTs had the opportunity to: (i) examine the complexity of enacting reform based practices that call into question beliefs about their teaching and learning; (ii) inquire and reflect on their practice by framing and reframing practice (Schön, 1983); (iii) learn from practice in ways that generalized to subsequent practice (Loughran, 2002); and (iv) create a knowledge of practice curriculum of novice practice for future PSTs (Cochran-Smith & Lytle, 2001).

Supporting new teachers to learn the understandings, skills, and sensibilities foundational to mathematics education reform and the teaching profession begins in university professional education. What is involved in this professional education is strongly debated (Cochran-Smith & Zeichner, 2005), but one thing that is common to many recommendations is the goal of cultivating PSTs' capacity to reflect on practice so that they may learn from it (Loughran, 2002). However, the means for supporting this development of reflection varies from one professional education program to another (Zeichner & Liston, 1996).

One means for developing PSTs' capacities to reflect on practice is using practice-based materials (cases, lesson plans, student work) and using the kinds of thinking cultivated using the materials to examine PSTs' practice. In this chapter I discuss the ways that I structured an assignment in which PSTs constructed a dilemma-based case on their *own* teaching (a problem of practice) and reflected on it. The assignment required PSTs to identify a problem of practice, construct a case that captures the problem, and reflect on the case using knowledge of teaching and learning. I developed the case assignment in response to my dissatisfaction with my capacity to cultivate PSTs' reflection on their practice in my methods class. Although my PSTs had been able to engage in thoughtful reflection when considering practice embedded in published cases and video (the practice of others), I struggled to engage them in the same kind of thinking on their own practice.

From my experience of working with my PSTs, I found that reflecting on *others'* practice was not sufficient for cultivating reflection on one's *own* practice. The nature of my PSTs' reflections on their practice, and their bewilderment when I pressed them to reflect, uncovered this difference. I needed my PSTs to both be able to reflect on their practice and to recognize the importance of this type of reflection as a means of learning from practice if they were going to take this skill into their careers as teachers (Loughran, 2002; Schön, 1983).

Vignette: Sharing Practice in Mathematics Methods Class

Gil exclaimed as he entered class, "half of my kids were absent today and you would not believe what happened! THE most quiet students raised their hands to answer my questions. I couldn't believe it!" Gil's class of 42 highly capable students had constructed math problems and shared solutions that day instead of using textbook tasks. Gil's colleagues responded to his surprise by relating how they got students to respond to questions. As I asked questions

about the mathematics, students' questions, and Gil's
instruction, I received quick responses or more stories from
the classroom that related to the event. After a few moments
the conversation waned and one PST summarized,
"Sometimes students just decide to talk. It's probably
because there was more time for them to talk with half of
them gone."

Gil's sharing of his surprising event was similar to many that
PSTs shared in methods class. My purpose for having them share
was for PSTs to use the ways of examining practice cultivated
using practice based materials to make sense of their practicum
experiences. Yet, what usually transpired was PSTs shared
stories of their practice with little detail, which were difficult for
others to question, and the group moved to drawing conclusions
based on little reasoning. Neither Gil, nor his colleagues, raised
issues examining multiple perspectives or considered
implications for their teaching. The discussion of Gil's surprising
event, and subsequent missed opportunities to make sense of his
practice, made me keenly aware that even though we had
developed skills to examine others' practice, the PSTs needed to
reflect on *their* problems of practice so that they could learn
from their teaching. However, as illustrated in the vignette
above, story sharing seemed to serve a different purpose for
PSTs than I had intended.

It seemed that for the PSTs, their purpose for sharing
surprising events and problems of practice focused on assuring
themselves that their teaching experiences were similar and
therefore normal. Certainly it is important for PSTs to feel
reassured when learning to teach. Yet, if they were going to learn
from *their* practice, they would need to learn how to reflect on
their practice. My dilemma was how to focus PSTs' inquiry into
their practice in ways that assured sharing of practice was an
opportunity to learn. As Loughran (2002) asserted, "It is [PSTs']
ability to analyze and make meaning from [their] experience that
matters most—as opposed to when the teacher educator ...
shares the knowledge with the [PSTs]" (p. 38). I had not done a
good enough job supporting the PSTs' reflecting on their

teaching. I needed ways to structure reflections so that PSTs were doing the analysis of their practice and learning from it.

Normalizing Problems of Practice and Then What?

Recognizing that I needed to learn how to support PSTs to do their own inquiry, I engaged in reading literature documenting when teacher communities pursued learning from practice. Little and Horn (2007), examining potential learning opportunities when teachers engaged in sharing their practice, described a phenomenon of normalizing experiences. Normalizing, or suggesting an event was "an expected part of classroom work," reassures teachers and cultivates solidarity among colleagues. Similarly, I found my PSTs normalized problems of practice by sharing stories as a means of reassuring each other that surprises and problems happen to all of them.

Normalizing was common in Little and Horn's examination of teachers' discussions, yet what was critical to productive discussions was the nature of the discourse after responses of normalizing. In my methods class, PSTs normalized Gil's surprise with their sharing of similar experiences, then they concluded that the reason the students' actions changed that day was solely based on something that was out of their control—student attendance. Little and Horn offer another path, one in which teaching becomes an "object of collective attention" and teachers engage in an intertwining process of: (1) normalizing practice; (2) inquiring into the specifics of practice (teaching, learning, and content to name a few); and (3) generating principles of teaching. This process was one I thought could be vital to developing the capacity of my PSTs to reflect on practice and learn from it.

Reflection as Inquiry into Practice

Little and Horn's work provides images of generative discussions of problems of practice. However, they are not explicit about *how* teachers might learn to dig into issues of practice. Scholars' considerations of productive reflection are

particularly insightful on what might be entailed by Little and
Horns' inquiring into the specifics of practice (Davis, 2006;
Loughran, 1996, 2002). Essential to notions of productive
reflection is Schön's (1983) idea of focusing on a problem of
practice with a process called framing and reframing (*e.g.*,
bringing to bear multiple perspectives on the issue).

Davis (2006), building from Loughran (2002) and Schön's
(1983) work, suggested that productive reflection for PSTs
requires practitioners to frame the problem considering the
learner, learning, subject matter, assessment, and instruction. The
use of multiple frames allows for viewing the problem from
different angles. It involves untangling the complexity of an
issue by reasoning about the facets of the problem and
considering potential resolutions that inform subsequent practice
(Schön, 1983). Davis' work suggests these frames for
considering practice are overlapping in expert practitioners'
thinking and in a teaching and learning context. As a result, she
advances that PSTs' productive reflection would connect or
integrate frames to make sense of the complexity of practice.
Loughran (2002), also connecting to Schön's (1983) work,
claimed that productive reflection involves learning that is
generative for subsequent practice.

To advance how a teacher educator may support PSTs
reflection I designed a specific assignment (to be described in the
next section) for my methods class and ask the following
research questions, informed by the work on reflection. (1) To
what degree do PSTs consider multiple frames when reflecting
on practice in their cases and reflections? (2) To what degree do
they integrate these frames to consider the complexity of
teaching and learning? (3) What opportunities to learn from
practice does the case assignment present for PSTs?

Background

The case assignment took place during the winter term of a
one-year professional education program and was based on
PSTs' teaching experiences in their fall practicum. Similar to
Schön's (1983) notion of "problem setting," my PSTs were

instructed to focus on a dilemma of practice that caused some sense of uncertainty, surprise, or presented a problem (Table 1). Because cases were used across their courses, the assignment connected to ways of capturing practice the PSTs had experienced. We collectively agreed that their cases should describe events that did not have clear-cut solutions, require consideration of multiple issues to make sense of the event, and provide insights on teaching and learning applicable beyond the specifics of the case (Carter, 1999; Shulman, 1992). Accompanying the assignment was a reflection on the case grounded in course work material and the conceptual framework of the professional education program.

Table 1
Case Assignment

Case Details
  The case should vividly focus on your dilemma, provide enough detail to see the issue, and not lose the reader in detail.
  The case should not solve the dilemma but elicit conversation.
  The narrative should include 2–4 questions to reflect on after reading the case.

Reflection Details
  Discuss the potential of the case to develop yours and your colleagues' understandings of the professional knowledge needed for teaching components: knowledge of content, knowledge of general pedagogy, knowledge of pedagogical content, and knowledge of context (Grossman, 1990; Shulman, 1986). How did you tap into your developing knowledge base in these areas to develop and think through this dilemma of practice?

Lee Shulman (1987) and colleagues' work on the professional knowledge base for teaching was central to the framework for the PSTs' professional educational program and their methods courses were steeped in ideas of reform practice. However, typical of students in teacher education, many of the PSTs understood learning to teach as a technical craft and came to their professional education program having considered only a direct instruction model of teaching (Grossman, 2005). About half of the PSTs were skeptical of the reform pedagogy described in text/video cases and articles. However, a few PSTs openly rejected their "apprenticeship of observation" in traditional school mathematics and stated that they were willing to try on these new ideas.

Data Sources and Analysis

Nine mathematics PSTs' cases and reflections were analyzed for this study. The analysis of PSTs' cases was completed in a number of phases. First, I read each case noting ideas to initially identify evidence for each frame. After reading and noting passages that related to the frames, I went back to examples of Davis' (2006) coding of data and compared, realigning my ideas about each frame as needed (Strauss, 1987). Then I returned to PSTs' cases, completed analytic summaries to reduce the data, and identified PSTs' central dilemmas. Finally, I catalogued areas in the text related to the professional knowledge base and frames ((i) learner and learning, (ii) subject matter, (iii) assessment, and (iv) instruction).

In my summaries I distinguished between ideas that were emphasized in a case, noting the amount of detail and development of the idea, and frames that were merely mentioned, but not developed. I also recorded what ideas were advanced about subsequent practice. To answer my three research questions I memoed, based on my analytic summaries, on the themes across the PSTs' cases, connections or lack of connection across frames, and ideas on subsequent practice (Miles & Huberman, 1994). Finally, I distilled the ideas into a table to compare across cases and identify patterns in the ways PSTs were using the frames, knowledge base and other issues that seemed central to the PSTs' reasoning that were outside of the frames or knowledge base. Some of these issues included consideration of equity and professionalism.

Many times PSTs' ideas could be related to more than one frame. For example, when a PST discussed issues of engaging students using manipulatives, he wrote about the mathematical ideas being considered, students' understandings using the tools, his assessment alignment with his instruction, and instructional decisions he made. This example was coded with learner/ learning, subject matter, assessment, and instruction. I also noted whether these ideas were simply considered as a sequence of ideas or linked to inform the PST's sense making.

Findings

All nine cases presented problems PSTs faced in their teaching. Six of the PSTs explicitly stated that they experienced a sense of surprise, frustration, or puzzlement, about the situation when it happened in their classroom. Three PSTs suggested that the assignment helped them uncover that there was not an "easy fix" for what had happened and that they needed to learn more about making sense of students' thinking and instructional strategies. Learning to teach was not just a technical craft but required deep consideration of teaching and learning. One PST suggested that originally he thought the problem he faced while teaching was a "simple, small" issue, but as he reflected on all that was at play he realized that his experience called into question his understandings and values presenting a "philosophical conundrum" (PST3).

Problems of Practice

Most striking in the cases was that all PSTs considered issues related to reform mathematics. Cases illustrated PSTs' struggles with reform pedagogies. For example, using group work for problem solving, asking students to share solutions, or using manipulatives to model mathematical relationships presented problems or a sense of uncertainty for PSTs.

Four of the nine PSTs explicitly discussed that their beliefs about teaching and mathematics were challenged as they reflected on their use of reform practices. One PST's experimenting with group work uncovered that he was "a beginning teacher, who [was] in the middle of the transition from traditional teaching to educational reform" (PST2, p. 2). Another PST's case writing exposed what he called "biases" about particular ways of instructing mathematics. "I had to get over my ... belief that manipulatives were a crutch...a less sophisticated way of engaging with mathematics" (PST4, p. 4). He also suggested how he had changed his ideas. "What I should be ... emphasizing that everything we did ... is completely legitimate in the mathematics world. My students are attacking problems in the same way a 'real' mathematician would" (p. 3). This PST examined his beliefs about mathematics. As result, his understanding of the nature of mathematics and teaching mathematics were considered and a new understanding of both emerged.

The case assignment was a vehicle for considering reform as more than a new menu of strategies to use in their teaching. PSTs were able to: reflect on their practice, consider multiple issues, and revise how they might approach their dilemma. Their work was evidence that they were inquiring into their practice. Ball (1997) noted one aspect of inquiry is that it "strive[s] to make a new idea viable, getting it on the table for examination, trial, and debate... not pushing it as 'the way' [or] 'selling it'" (p. 94). PSTs' writing weighed the benefits and drawbacks in light of a variety of sources of information. As a result, no one made a "sales pitch" in their case suggesting that they had the "silver bullet," nor was reform the magical hope for student learning. PSTs played out their thinking, albeit at varying levels of complexity, to examine their beliefs and understandings of teaching and learning mathematics.

Using Multiple Frames to Analyze Cases

To address my first two research questions on PSTs' use of frames to reflect on practice, I found that PSTs' cases were grounded in complex issues, and as a result most PSTs shared how they wrestled with student learning, content, instruction, and other issues. The reflection that followed the case allowed the PSTs to unpack the case and to analyze what was at play in the problem. Table 2 summarizes the topics addressed and the frames considered in the cases.

PSTs used a number of frames to make sense of their dilemmas of practice (Table 2). The most common frames were learner/learning and instruction. Unlike the nature of discussions of practice in methods class, it is evident that PSTs considered a number of perspectives in their reflections on practice. Even though not all of these ideas were integrated, which Davis (2006) considers critical to productive reflection, seven of the nine PSTs had some degree of integration either in the case or the reflection. Integrating frames in reflections resulted in a number of PSTs being able to consider affordances and drawbacks of instructional moves and potential changes. The case assignment structured PSTs reflection on practice such that they had opportunities to share their practice and learn from it.

Table 2

PSTs Dilemmas of Practice and Frame for Analyzing Dilemma

	Cases of:	Frames in Case & Reflection
1	Teaching math procedures versus reasoning (how vs why), learner needs both.	Learner/Learning, Subject Matter, Instruction, Pedagogical Content Knowledge, Professionalism
2	Problem solving in heterogeneous groups – high and low students working together.	Learner/Learning, Instruction, Pedagogical Knowledge
3	Cultivating mathematical discourse – individual needs and whole class learning	Learner/Learning, Instruction, Pedagogical knowledge, Equity and Special Needs Students
4	Using Algebra tiles to solve equations—valuing multiple representations in class and assessment.	Learner/Learning, Subject Matter, Assessment, Instruction, Knowledge of Content, Pedagogy, and Pedagogical Content

	Cases of:	Frames in Case & Reflection
5	Teaching without *just* showing procedures—using Algebra tiles as means to manipulate equations.	Instruction, Knowledge of Context, Other Teacher's Instruction
6	Maintaining cognitive challenge and modifications for ELL students	Learner/Learning, Assessment, Instruction, Knowledge of Pedagogy and Context, Equity, Professionalism,
7	Meeting students' needs and engaging them in contextual learning	Learner/Learning, Subject Matter, Assessment, Instruction, Knowledge of Content and Pedagogy
8	Using students' solution sharing to guide whole class discussion	Learner/Learning, Instruction, Knowledge of Pedagogy and Context, Professionalism
9	Teaching conceptually—shifting students' practices and expectations	Learner/Learning, Subject Matter, Instruction, Knowledge of Pedagogy and Pedagogical Content

As I analyzed cases, I saw that many PSTs were healthy skeptics of reform and had seen the value of reflecting on practice. However, not all PSTs came to understand *how* to examine practice from multiple perspectives. A few PSTs' reflections considered a limited number of perspectives, were vague in attempts to frame their dilemmas, or discussed frames as a sequence of considerations rather than considering how perspectives interacted. One PST's case focused on a new instructional practice and was framed by a perspective outside the author's locus of control. This framing resulted in drawing conclusions about students' performance based on a narrow perspective. Perhaps more importantly, the PST's case lacked specificity and asserted certainty in ways that made it difficult for others to suggest reframing the case.

Reflecting on Practice to Learn from Practice

The main aim of reflection, as advanced by Schön (1983), is to learn from practice by problem solving on events of teaching. PSTs' cases and reflections enacted what Schön calls "reflection on action," a chance to look back on practice without the stress of in-the-moment decision-making. To address my third research question, what opportunities to learn from practice did the case assignment present for PSTs, a number of PSTs directly talked about their opportunities to learn based on the assignment. In fact, five of the nine PSTs' cases and reflections cited explicitly that their study of practice provided a chance to think more deeply about the problem, to consider "what I value, what I am skilled at, and what I need to work on," (PST3) to think more about "specific instances... [instead] of a lesson as a whole" (PST9), to do more than just remember what they did, but to analyze their practice.

A number of PSTs juxtaposed the written reflections on lessons constructed immediately after their teaching to the case assignment. One PST noted, "for my [reflection on the day of the lesson] I didn't really step-back and think about what I could have done differently from the beginning. I was already thinking about what I was going to do for my next lesson to make up for lost time" (PST7, p. 3). A number of PSTs suggested that the time and distance away from the immediate "what will I do next" type questions and the chance to consider different perspectives and issues were crucial to their learning. PSTs drew on text presented in their courses and the input of peers reviewing drafts of their cases. One PST noted that when he considered different sources for framing his case he was faced with inconsistent answers to problems of practice. His case and reflection brought the uncertainty of teaching and learning to light and required him to weigh the ideas. He noted that he had to "learn how to balance these inconsistencies to build [his] teaching style" (PST2, p. 3). The PSTs seemed to realize the purpose of reflecting on *their* practice, not just for storytelling, rather as an opportunity to learn from their practice.

Reflection on Practice to Inform Future Practice

Reflection on action supported all PSTs to think about their future teaching. Each PST's reflection made some reference to future practice or changes that the PST would make in subsequent teaching. However, and perhaps more important, was the fact that some PSTs were able to think more critically about their practice, to abstract and see the dilemma in the case not just as an instance in their teaching, but to consider what was to be learned in terms of generalizable ideas. PSTs grappled to understand the issues at play when trying on reform. The process of reflecting pressed them to abstract from the case so that they had a repertoire of responses ready. This type of thinking was observed especially when PSTs' reflections integrated frames to consider the complexity of practice. It seems that in this deliberation of multiple frames, PSTs were able to specify details of practice and emerge with what Little and Horn (2007) consider principles of practice. This was most notably captured in PST4's reflection when he discussed his biases uncovered by reflecting on his use of manipulatives. PST4 saw his prior perspectives on teaching and learning mathematics as potentially limiting access for student learning. He advances a principle that all students have the right to learn and it is a teacher's responsibility to not limit these opportunities. Another PST's reflection exposed his principle that teaching should attend to why mathematics works and why it is important as a means to construct "higher quality mathematics education for students in my class" (PST1, p. 3). PSTs' case assignments supported PSTs discovering principles that were tacit in their thinking and made the principles explicit so that their future practice might benefit.

Implications for Cases as Knowledge-of-Practice in University Professional Education

The PSTs' case assignment presented the opportunity to learn from practice and uncovered problems of practice situated in trying on reform based practices. As a whole, the PSTs were able to frame their problems of practice in complex interactions using multiple frames. However, from this one assignment I am not suggesting that PSTs will always be able to reflect on practice to learn from it. In fact, my analysis uncovered that a few PSTs, to varying degrees, saw the assignment as a chance to narrowly examine their problem and assert with certainty what was at play in their practice. To cultivate and sustain PSTs' capacity to reflect on practice they need more opportunities to come to see that they could frame their problems of practice in multiple ways. As a teacher educator, I could easily frame PSTs' practice by commenting on their reflections; however, I would be falling prey to Loughran's (2002) warning that I raised at the beginning of this chapter. I would be the one inquiring into PSTs' practice rather than the PST.

Cochran-Smith and Lytle's (2001) work from almost a decade ago, and Little and Horn's (2007) recent work on teacher learning communities, both suggest the potential of teachers learning from practice. In their work, learning from practice becomes a knowledge base or resource from which teachers draw. Similarly, the PSTs' cases could become a knowledge base from which PSTs can learn. To further explore the utility of this assignment I plan to employ the PSTs' cases each year as a part of the university professional education curriculum. Using PSTs' cases and carefully attending to future PSTs' analysis of these cases will support a knowledge-of-practice (Cochran-Smith & Lytle, 2001). As this way of learning from practice becomes more normative this type of reflection on practice may disrupt the conventional socialization to teaching and thus change what teachers talk and think about to develop a narrative of inquiry (Ball & Cohen, 1999).

References

Ball, D. L. (1997). What do students know?: Facing challenges of distance, context, and desire in trying to hear children. In B. J. Biddle, T. L. Good & I. Goodson (Eds.), *International handbook of teachers and teaching* (pp. 769–818). Dordrecht, The Netherlands: Kluwer Publishing.

Ball, D. L., & Cohen, D. K. (1999). Developing practice, developing practitioners: Toward a practice-based theory of professional education. In L. Darling-Hammond & G. Sykes (Eds.), *Teaching as the learning profession: Handbook of policy and practice* (pp. 3–32). San Francisco, CA: Jossey Bass.

Carter, K. (1999). What is a case? What is not a case? In M. A. Lundeburg, B. B. Levin & H. L. Harrington (Eds.), *Who learns what from cases and how?: The research base for teaching and learning with cases* (pp. 165–175). Mahwah, NJ: Lawrence Erlbaum.

Cochran-Smith, M., & Lytle, S. L. (2001). Beyond certainty: Taking an inquiry stance on practice. In A. Liberman & L. Miller (Eds.), *Teachers caught in the action: Professional development that matters* (pp. 45–58). New York: Teachers College Press.

Cochran-Smith, M., & Zeichner, K. M. (2005). *Studying teacher education: The report of the American Educational Research Association panel on research and teacher education.* Mahwah, NJ: Lawrence Erlbaum.

Davis, E. A. (2006). Characterizing productive reflection among preservice elementary teachers: Seeing what matters. *Teaching and Teacher Education, 22*(3), 281–301.

Grossman, P. (1990). *The making of a teacher: Teacher knowledge & teacher education.* New York: Teachers College Press.

Grossman, P. (2005). Research on pedagogical approaches. In M. Cochran-Smith & K. M. Zeichner (Eds.), *Studying teacher education: The report of the American Educational Research Association panel on research and teacher education* (pp. 425–476). Mahwah, NJ: Lawrence Erlbaum.

Little, J. W., & Horn, I. S. (2007). Normalizing problems of practice: converting routine conversations into a resource for learning in professional communities. In L. Stoll & K. S Louis (Eds.), *Professional learning communities: Divergence, depth, and dilemmas* (pp. 79–92). New York: Open University Press.

Loughran, J. (1996). *Developing reflective practice.* London: Routledge-Falmer.

Loughran, J. (2002). Effective reflective practice: In search of meaning in learning about teaching. *Journal of Teacher Education, 53*(1), 33–43.

Miles, M. B., & Huberman, A. M. (1994). *Qualitative data analysis* (2nd ed.). Thousand Oaks, CA: Sage.

Schön, D. A. (1983). *The reflective practitioner: How professionals think in action.* New York: Basic Books.

Shulman, L. S. (1986). Those who understand: Knowledge growth in teaching. *Educational Researcher, 15*(2), 4–14.

Shulman, L. S. (1987). Knowledge and teaching: Foundations of the new reform. *Harvard Educational Review, 57*(1), 1–22.

Shulman, L. S. (1992). Toward a pedagogy of cases. In J. H. Shulman (Ed.), *Cases methods in teacher education* (pp. 1–30). New York: Teachers College Press.

Strauss, A. (1987). *Qualitative analysis for the social scientist.* Cambridge: Cambridge University Press.

Zeichner, K. M., & Liston, D. P. (1996). *Reflective teaching an introduction.* Mahwah, NJ: Lawrence Erlbaum.

Endnote

1. This work was supported in part by funding from Oregon State University's Center for Teaching and Learning Innovations Grant. The opinions expressed here are the authors and do not reflect the views of the funding agent.

Rebekah Elliott is an assistant professor of mathematics education at Oregon State University. Her research and teaching interests include content and pedagogies needed for teaching, learning to teach mathematics, and learning to lead professional education.

Van Zoest, L. R. and Stockero, S. L.
AMTE Monograph 6
Scholarly Practices and Inquiry in the Preparation of Mathematics Teachers
© 2009, pp. 153–169

10

Deliberate Preparation of Prospective Teachers for Early Field Experiences

Laura R. Van Zoest
Western Michigan University

Shari L. Stockero
Michigan Technological University

Our ongoing efforts to develop an early mathematics field experience that alleviates some well-established limitations of the traditional apprenticeship approach have highlighted the importance of deliberately preparing prospective teachers to work with students. Here we describe some key preparation efforts— engaging with the mathematical task to be used with students, analysis of student thinking, structured planning, and purposeful reflection—and then discuss the prospective teachers' perceptions of the effectiveness of these activities in helping them meet the goals of the field experience.

Although early field experiences have long been advocated for as a means to initiate prospective teachers into mathematics teaching (Ishler & Kay, 1981), their limitations have also been identified through time (e.g. Dewey, 1904/2008; Zeichner, 1980). Over a century ago, Dewey proposed the *laboratory* approach as an alternative to the common practice—then and now—of an *apprenticeship* approach to early field experiences. Documented to enhance prospective teachers' learning (Philipp et al., 2007), the laboratory approach shifts the focus from *replicating* what teachers are observed doing in the classroom to

153

thinking about teaching and learning and making connections between theory and practice. Consistent with this, Darling-Hammond and colleagues (2005) report that successful teacher education programs include carefully designed early field experiences that focus prospective teachers' attention on student learning and its relationship to teacher actions. Movement away from an apprenticeship approach seems particularly relevant when the way prospective teachers are being asked to teach (e.g. National Council of Teachers of Mathematics, 2000) is dramatically different than the instruction they likely experienced as students (Stigler & Hiebert, 1997).

In our ongoing work to design and implement experiences consistent with a laboratory approach (e.g. Van Zoest, 2004; Van Zoest & Stockero, 2008a, 2008b), it has become increasingly clear to us that while the design of the school-based experience is important, additional work is needed to maximize the benefits of early field experiences. In particular, prospective teachers must be deliberately prepared and positioned to learn from their experiences. In the following, we begin by providing a brief description of a course-based field experience. We then introduce course activities deliberately designed to enhance our prospective teachers' experiences in the field by focusing them on things they need to think about as they interact with students and ways to use the ideas of the course in doing so. To conclude, we discuss the prospective teachers' perspectives on the extent to which these planned activities supported them in meeting the goals of the field experience.

Field Experience

The field experience consists of three methods class sessions, each 110 minutes in length, spent in local middle school classrooms. During these site visits, the prospective teachers (PTs) work with groups of 3–5 students to solve a mathematics problem that they have worked with extensively in the methods course. The goal for this common experience is not for the PTs to "teach" students how to solve the problem by telling or showing them how to do it. Instead, following Aiken and Day's (1999) recommendation to provide specific goals, we explicitly state that their goal as a teacher is "to practice listening to, assessing, and building on students' thinking for the purpose of teaching mathematics." Key features of the field experience— a preselected task, an explicitly stated teaching goal, use of a limited number of classrooms, and working with a small group for a designated length of time—allow a degree of control over the experience that is not present when individual PTs are placed in individual teachers' classrooms. It also provides the PTs with important, but manageable components of teaching with which to engage (e.g. Hiebert, Morris, & Glass, 2003).

During each school visit, the PTs work in teacher-documenter pairs, each taking on the role of teacher for one middle school class period. The documenter records successes and missed opportunities in three areas: (1) giving the right amount of information; (2) pushing students to think harder; and (3) checking student understanding. The documenter's observations, an audio recording of the session, and the middle school students' written work are used as evidence for a post-teaching reflection paper.

The field experiences are intentionally sequenced to provide two important opportunities for learning: planning to use two different problems with the same group of students and planning to use the same problem with two groups of students. The first allows the PTs to use their knowledge of their students to think about how to teach a different lesson in a way that is responsive to their needs, and the second to use their experience with the lesson to rethink how it could be better facilitated. To accomplish this, we use two different mathematical tasks and two different classrooms. The sequence of the problems and classrooms is shown in Figure 1.

Field Experience 1	Field Experience 2	Field Experience 3
Problem A	Problem B	Problem B
Classroom 1	Classroom 1	Classroom 2

Figure 1. Problem and classroom sequence.

Preparation

We engage the PTs in iterative preparing-enacting-reflecting cycles that include extensive pre-field experience activities, time spent in middle school classrooms, and structured reflection that helps them plan for subsequent classroom visits. Similar cycles have been advocated for and used by teacher educators in a variety of contexts because of their potential to help PTs make connections between theory and practice, learn to adapt their teaching to varying groups of students, and consider how their practice could be improved (e.g. Artzt, 1999; Hiebert et al., 2003; Lee, 2005). Through ongoing analysis of our work with PTs, we have identified a number of productive activities for them to engage in as part of the preparation phase of this cycle. These include: investigating their own and others' mathematical thinking related to the tasks they will use with students; practicing listening to, analyzing and responding to student thinking; and considering examples of reflection used as a tool for purposeful improvement of instruction. In the remainder of this section we elaborate on these activities and discuss how they support the PTs in their early field experiences.

Mathematical Thinking

Indirect field experience preparation began on the first day of class when PTs were challenged to solve a mathematics problem in multiple ways, present and discuss their solutions, and analyze their peers' solutions. We, as well as others (e.g. Hollebrands, Wilson, & Lee, 2007; Lampert & Ball, 1998), have found this to be an important first step in planning to teach for student understanding. Many PTs have a limited conception of what constitutes a correct or potentially fruitful mathematical solution (Bowers & Doerr, 2001), are uncomfortable with solution strategies other than their own (Lee, 2005), and only take note of responses that are aligned with what they expect to hear, dismissing novel or unexpected ideas. Thus, opportunities to consider alternate strategies are an important prerequisite to predicting ways that middle school students might think about a problem and considering how they might respond to various solution methods—activities the PTs will be asked to do later in their field experience preparation. In particular, we find that engaging in mathematics in this way stretches PTs' thinking about the mathematics in the problems and pushes them to begin to justify their mathematical thinking, rather than simply report how they solved a problem. For example, most have not considered visual approaches to solving problems. Even when given a picture that describes a mathematical situation, their tendency is to abstract the mathematics and ignore the corresponding picture.

The second day of class built on this experience as the PTs watched video[1] of students engaging with the same problem, discussed ways the students thought about the problem, and related the students' solutions to their own or classmates' solutions. As others (e.g. Lampert & Ball, 1998; Sherin & van Es, 2005) have suggested should be the case, these video clips focused on students, not teachers, which provided an opportunity to push the PTs to analyze student thinking in a way that can inform instruction—focusing on what students understood rather than simply on what they did or said. In addition, the videos serve as a means for them to begin thinking about the relationship between teacher actions and student learning—particularly the kind of student thinking teachers can evoke, how they evoke it, and how they use it in the lesson—and to make connections between the ideas in course readings and teaching practice. The cycle of engaging with the mathematics and then analyzing students' understanding of the same mathematics was repeated with several high quality tasks, two of which were later used for the field experience.

Another important part of the early preparation was a three-column Prediction Chart with the headings "Learning Activity," "Potential Student Reactions," and "Your Response." Immediately after completing the above cycle for each field experience problem, the PTs drew on their experiences and course readings to list specific actions they would take to engage students with the problem, mathematical ways in which students might react to these actions, and how the teacher might respond in turn—new questions to ask and strategies for engaging

students while maintaining what Stein and Smith (1998) have
called the cognitive level of the task.

Preparing for the Classroom

Setting the context. Direct field experience preparation began
approximately two weeks prior to the first school visit. The PTs
were introduced to the context and nature of the field experience
using video of a PT working with a group of four students during
a previous semester. This 5-minute splice of video was chosen
because it included four distinct ways of thinking—one from
each student. The focus of watching the video was on the
students' thinking about the mathematics and the ways in which
the PT contributed to or could have better supported that
thinking. Discussing this video provided more practice in
making sense of students' thinking and alleviated some of the
anxiety caused by not knowing what to expect in what, for many
of them, was their first visit to a school as a PT. This also
provided an opportunity to consider facilitating the discussion of
a small group of students, as the other videos watched in class all
involved whole-class situations.

Structuring the lesson. Once the context had been set, the
PTs worked in teacher-documenter pairs to develop a joint
lesson plan. Taking into account both that most of the PTs had
not previously written a lesson plan and the recommendation that
teaching situations be posed in "meaningful but digestible
chunks" (Hiebert et al., 2003, p. 205), we provided two key types
of structure to the plans that we have found to help PTs learn
from their field experience. First, we provided the lesson goals
and a mathematical problem on which to center the lesson. This
reflects our attempt to focus PTs on planning their actions in
response to student thinking rather than on deciding what to
teach. The learning goals focused the PTs on the dual goals of
the field experience—to provide a positive learning experience
for students and to practice accessing, analyzing, and building on
students' thinking to support them in developing mathematical
understanding.

Second, we provided a *launch, students-at-work, summary*
structure for the lesson itself—reflecting what might be

encountered in reform-based curricula materials—and pushed them to think about the role each of these components plays in developing student understanding about mathematics. We began with the students-at-work section because we wanted the PTs to first have a strong idea of what they would be asking the students to do and think about so they would then have a better idea both of what they didn't want to give away in the launch and for what they wanted to use the launch to set the stage.

The PTs began the students-at-work section by revisiting their Prediction Charts, analyzing which aspects supported the lesson goals, and then incorporating or modifying their original work accordingly. Similar to the planning process described by Hiebert and his colleagues (2003), the PT's development of the students-at-work section was conceived as a way for them to anticipate students' thinking and develop possible responses so that they would be better prepared to react to their group of middle school students' thinking in-the-moment and facilitate a discussion around the mathematics in the task.

After the PTs drafted the students-at-work section, we turned their attention to the launch and summary and how these sections could support what they wanted to accomplish in the lesson. To prompt them to think deeply about the launch, we used a video of a teacher launching a similar problem as the basis for analyzing the components of an effective launch. The emphasis for the summary planning was on thinking about what they wanted students to take away from the lesson and what they would do to support that. We have observed the summary to often be the weakest part of experienced teachers' lessons and it is equally, if not more, difficult for the PTs. Thus, we prompt them to be specific about what they will do and say to bring closure to the lesson and, during the classroom visits, cue them when the end of the period is approaching to allow time for them to complete the summary phase of their lesson—a deliberate attempt to develop good teaching habits.

We also use the planning stage of the cycle as an opportunity to support the idea that a lesson plan is a work in progress to be reflected on and revised both before and after implementation. To help with this, after drafting their plan, the PTs were provided with the following questions to ask themselves: "Have you focused on the students' mathematical thinking? Have you provided enough detail about what you will have the students do, questions you will ask the students, and aspects of the students' reactions you will focus on? Who is doing the bulk of the talking? The thinking?" Next, their modified plan was critiqued by other teacher-documenter pairs, using a format that asked them to identify "something liked" and "something wondered about." The PTs were instructed to look for trends in this peer feedback and incorporate their responses into their final lesson plan.

Purposeful Improvement

The *same students–different problem* and *different students– same problem* model allowed for preparation to continue throughout the field experience. It also added richness to whole-class debriefing sessions that took place throughout the field experiences; these focused on comparing outcomes of teaching sessions in relation to how the teacher facilitated them. This debriefing in community seemed critical, as it provided more examples of student thinking and opportunities for PTs to be held accountable for their conclusions and to consider alternate interpretations of events. To ground the debriefing, we analyzed selections of transcribed dialogue from PTs' teaching sessions offered by the PTs and chosen by the instructor for their likelihood to prompt productive discussion. The debriefing conversations paralleled those around the videos, focusing on student thinking and how the teacher either supported or inhibited such thinking.

Following the first classroom visit, the PTs worked with their partner to reflect on their teaching sessions. This paired reflection was intended as an opportunity for the pairs to discuss their interpretations of the teaching sessions and share suggestions about how their joint lesson plan could be improved. Each pair produced a paper in which they: (1) used Stein and Smith's (1998) Cognitive Demand Framework as a tool to assess how their joint lesson plan supported their ability to maintain high-level cognitive demand during the lesson; (2) chose two students (from either PT's small-group session) who thought about the problem differently and compared and contrasted their thinking; and (3) identified specific instances where either partner helped or hindered students' development of mathematical understanding. As described by Artzt (1999), intentional structuring of reflection provides a tool for improving instruction. Our structure focused the PTs on student thinking, and the relationship between the development of this thinking and their actions as a teacher.

Reflection papers that addressed the same three prompts were completed individually following the second and third field experiences. In each case, the PTs were required to use transcripts, student work, and documenter notes to support their analyses. Substantial educative feedback was provided on each paper and the accompanying transcripts. In particular, the instructor used knowledge of the context, students, and mathematics to raise issues that the PTs may not have considered and to ask questions to prompt further reflection. In the words of one student, "The feedback in this situation never said this is wrong, and this works, but gave us questions to consider and ideas to pursue" (S4, F07). This was important for two reasons: 1) it gave the PTs something to think about as they began the next iteration of the planning-enacting-reflecting cycle; and 2) it modeled for them the types of questions reflective teachers ask themselves on a regular basis.

The Students' Perspective

During the fall 2007 and spring 2008 semesters, PTs enrolled in the methods course completed an online, anonymous survey (n = 15 out of 20 enrolled and n = 10 out of 12, respectively) about the extent to which different course activities supported them in meeting the goals of the course field experience. They rated each activity using a five-point scale that ranged from "not at all helpful" to "very helpful" and provided open-ended feedback. On the last day of each semester, the PTs (n = 20 and n = 12) wrote a confidential reflection on what they had learned from the course. These feedback sources are used to better understand the PTs' perceptions of our efforts to prepare them for the field experience.

The PTs were unanimous in their view that the *same students-different problem* and *different students-same problem* format was useful in supporting their field experience preparation: 78% considered teaching two different groups of students "very helpful" in meeting the goals of the field experience, with 100% responding "very helpful" or "somewhat helpful." The numbers were similar for using two problems with the same students (74% and 100%, respectively). The following is a representative explanation:

> Working with the students a second time helped me prepare what to do to help them find success when I met with them again. The same is true of presenting them with a different problem. By having worked with them before I was able to think how they would think when they saw this problem and how I needed to react. (S4, F07)

The PTs were less uniform in their reaction to the paired reflection. In the best cases, this model provided them with two perspectives on the experience and the opportunity to compare the student thinking that emerged from two sessions:

> Working with a partner on the first field experience reflection was very helpful. It was a good way for both of us to analyze what happened in the field experience. We were able to discuss things that we felt worked good and things that didn't work so well. I liked the idea of having someone else's opinion about what happened in the lessons. (S8, Sp08)

On the other hand, there were a number of PTs who had a negative reaction to being required to work with a partner. The following captures the tensions experienced by some:

> Working with a partner on the first field experience
> reflection, I believe, hindered me in analyzing what
> occurred because my partner did not reflect in depth
> about the experience and I was unwilling to redo what he
> had worked on. In working with my partner, we would
> have been at odds because I was unwilling to just get the
> project done. (S3, F07)

Thus, while we believe the paired reflection has the potential to
support more critical reflection on the field experiences, and thus
better planning for subsequent classroom visits, careful
consideration needs to be given to structuring it in a way that
helps, not hinders, each of the PTs.

When asked how doing and discussing the mathematics
supported their field experience work, 96% of PTs reported that
doing the mathematics was very (74%) or somewhat helpful
(22%), while 98% said the same of discussing the mathematics
(76% and 22%, respectively). The following is representative:

> Discussing the mathematics problems as a class was the
> most important aspect in preparing me for the field
> experiences. By doing this, I saw the ways my fellow
> classmates were thinking about the problems, which was
> helpful because this prepared me for the possible
> solutions that my students could come up with. I would
> not have been as comfortable discussing the solutions
> and problems with my students during the field
> experiences if we did not discuss them as a class first.
> (S2, Sp08)

Despite the fact that our main reason for engaging the PTs
with the mathematics was to prepare them for watching students
engage with the mathematics via video, engaging with the video
was not perceived by the PTs as being as valuable as doing and
discussing the mathematics itself. In the survey, only 14%
reported that watching the video was very helpful in supporting
them to meet the goals of the field experiences, although another
78% reported it was somewhat helpful. For discussing the

videos, 40% reported that it was very helpful (28% somewhat helpful). The latter was one of the lowest percentages of positive responses for any of the field preparation activities surveyed. Reasons given by PTs included, "the videos rarely introduced new thinking that we had not already thought of in class" (S7, F07) and "watching the videos was too quick-moving for me to get everything I wanted out of them" (S4, F07). Several PTs reported that they tended to focus on the teacher, which distracted them from analyzing student thinking.

We asked the students to assess the supportiveness of the reflective papers along four dimensions: (1) reflecting on the lesson plan (32% very helpful; 76% somewhat helpful or very helpful); (2) using the Cognitive Demand Framework (12%; 68%); (3) reflecting on student thinking (58%; 92%); and (4) reflecting on teaching (82%; 92%). The high ratings were impressive given the amount of work the papers demanded of the PTs—transcribing at least six minutes of audiotape, reflecting, analyzing, synthesizing, and writing—often in less than a week so that the instructor could provide feedback prior to the next visit. The lower ratings on the first two dimensions and the fact that both were rated by 20% of the PTs as not supporting the field experience are consistent with the sense we had that the PTs struggled to make the connection between their lesson plans and the level of thinking in which they were able to engage their students. This wasn't directly addressed by our field experience preparation activities, as we did not have plans for the lessons we watched on video. One change that we intend to make as a result of the feedback is to introduce the Cognitive Demand Framework earlier in the course and have the PTs use it as they reflect on the videos. Although they won't have access to teacher's actual lesson plans, they can analyze what happens during the lesson and consider things that one might plan in advance to maintain a high level of cognitive demand.

Conclusion

Much of what is often unproductive about early teaching experiences can be mitigated by carefully designing field experiences using a laboratory approach (Dewey, 1904/2008; Philipp et al., 2007) and deliberate preparation of the type we describe here. Elements of the deliberate preparation for our early field experience—specific mathematical tasks, explicit goals, an intense focus on mathematical thinking, structured planning that includes an iterative process of lesson improvement, and group reflection and abstraction based on common experiences—have allowed our PTs to begin to see the connections between theory and practice and learn to reflect on teaching in a structured way.

Despite this progress, there are issues that remain. For example, the idea that students think about mathematics in multiple ways is one that our PTs consistently take away from the experience, and one that we feel is particularly important in teaching for student understanding. However, comments such as the following suggest there is more work to be done:

> I learned the importance of accepting all methods in a classroom. For example, one student's method could make a problem easier for another student that couldn't understand the way I was teaching. (S2856, F07)

Although this PT recognized the value of multiple methods, the reason he or she gave—that a student might understand another student's method better than the teacher's—suggests that the PT has not understood the main purpose of multiple methods—to better understand the *mathematical ideas* that are being taught, rather than having different ways to solve a problem. As Bowers and Doerr (2001) suggest, a lack of this understanding can actually prevent teachers from trying to understand student thinking in order to build on incorrect, but potentially productive thinking. This, as well as other reflections, such as that below, point to the complicated nature and difficulty of promoting change:

> One main topic I'll be walking out of this course with is that not everyone learns best like I do ... before this class I had a concrete idea how I wanted to teach my

future classes and it was only based on my learning
tactics. Now I don't have a clue of how exactly I am
going to set up my future course, but at least I know now
not to leave those other students who don't learn like I
do in the dust (S1137, F07).

Thus, while we feel that in many cases we make progress in
addressing unproductive ideas about teaching, we agree with
others (e.g. Hiebert, Morris & Glass, 2003) that it isn't
reasonable to expect that PTs will have learned enough from one
class, one field experience, or even from a whole teacher
education program, to step into a classroom as a fully-formed
teacher. Yet, we must figure out how to use early field
experiences as a way to support the ideas in our coursework in
order to give PTs a foundation to build upon as they continue
their development as a teacher. At this point, we have settled on
student thinking as the best place to center our efforts. In the
words of one PT, "Our teaching needs to be flexible and a
reaction to student thinking, so to teach us how to teach, we need
to be taught to understand students" (0321, Sp08). Better
understanding the ways in which PTs respond to our efforts to
deliberately prepare them to work with students is critical as we
continue to revise our course activities in support of producing
teachers prepared to respond to calls for reform.

References

Aiken, I. P., & Day, B. D. (1999). Early field experiences in
    preservice teacher education: Research and student
    perspectives *Action in Teacher Education, 21*(3), 7–12.
Artzt, A. F. (1999). A structure to enable preservice teachers of
    mathematics to reflect on their teaching. *Journal of
    Mathematics Teacher Education, 2*(2), 143–166.
Bowers, J., & Doerr, H. M. (2001). An analysis of prospective
    teachers' dual roles in understanding the mathematics of
    change: Eliciting growth with technology. *Journal of
    Mathematics Teacher Education, 4*(2), 115–137.

Darling-Hammond, L., Hammerness, K., Grossman, P., Rust, F., & Shulman, L. (2005). The design of teacher education programs. In L. Darling-Hammond & J. Bransford (Eds.), *Preparing teachers for a changing world* (pp. 390–441). San Francisco, CA: Jossey-Bass.

Dewey, J. (1904/2008). The relation of theory to practice in education. In M. Cochran-Smith, S. Feiman-Nemser, D. J. McIntyre & K. E. Demers (Eds.), *Handbook of research on teacher education: Enduring questions in changing contexts* (3rd ed.). New York: Routledge.

Hiebert, J., Morris, A. K., & Glass, B. (2003). Learning to learn to teach: An "experiment" model for teaching and teacher preparation in mathematics. *Journal of Mathematics Teacher Education, 6*(3), 201–222.

Hollebrands, K. F., Wilson, P. H., & Lee, H. S. (2007). *Prospective teachers use of a videocase to examine students' work when solving mathematical tasks using technology.* Paper presented at the Twenty-Ninth Annual Meeting of the Psychology of Mathematics Education-North American Chapter, Reno, Nevada.

Ishler, P., & Kay, R. (1981). A survey of institutional practice. In C. Webb, N. Gehre, P. Ishler & A. Mendoza (Eds.), *Exploratory field experiences in teacher education* (pp. 15–22). Washington, DC: Association of Teacher Educators.

Lampert, M., & Ball, D. (1998). *Teaching, multimedia and mathematics: Investigations of real practice.* NY: Teachers College Press.

Lee, H. S. (2005). Facilitating students' problem solving: Prospective teachers' learning trajectory in a technological context. *Journal of Mathematics Teacher Education, 8*(3), 223–254.

National Council of Teachers of Mathematics. (2000). *Principles and standards for school mathematics.* Reston, VA: National Council of Teachers of Mathematics.

Philipp, R., Ambrose, R., Lamb, L., Sowder, J., Schappelle, B., Sowder, L., et al. (2007). Effects of early field experiences on the mathematical content knowledge and beliefs of prospective elementary school teachers: An experimental study. *Journal for Research in Mathematics Education, 38*(5), 438–476.

Seago, N., Mumme, J., & Branca, N. (2004). *Learning and teaching linear functions: Video cases for mathematics professional development* (pp. 6–10). Portsmouth, NH: Heinemann.

Sherin, M. G., & van Es, E. A. (2005). Using video to support teachers' ability to notice classroom interactions. *Journal of Technology and Teacher Education, 13*(3), 475–591.

Stein, M. K., & Smith, M. S. (1998). Mathematical tasks as a framework for reflection: From research to practice. *Mathematics Teaching in the Middle School, 3*(5), 268–275.

Stigler, J. W., & Hiebert, J. (1997). Understanding and improving classroom mathematics instruction: An overview of the TIMMS video study. *Phi Delta Kappan, 79*(1), 14–21.

Van Zoest, L. R. (2004). Preparing for the future: An early field experience that focuses on students' thinking. In D. R. Thompson & T. Watanabe (Eds.), *AMTE Monograph 1: The Work of Mathematics Teacher Educators: Continuing the Conversation* (pp. 124–140).

Van Zoest, L. R., & Stockero, S. L. (2008a). Using a video-case curriculum to develop preservice teachers' knowledge and skills. In M. S. Smith & S. N. Friel (Eds.), *AMTE Monograph 4: Cases in Mathematics Teacher Education: Tools for Developing Knowledge Needed for Teaching* (pp. 117–132).

Van Zoest, L. R., & Stockero, S. L. (2008b). Concentric task sequences: A model for advancing instruction based on student thinking. In F. Arbaugh & P. M. Taylor (Eds.), *AMTE Monograph 5: Inquiry into Mathematics Teacher Education* (pp. 47–58).

Zeichner, K. M. (1980). Myths and realities: Field-based experiences in preservice teacher education. *Journal of Teacher Education, 31*(6), 45–55.

Endnotes

[1] Both the mathematical tasks and the video clips were taken from the *Learning and Teaching Linear Functions* professional development curriculum (Seago, Mumme, & Branca, 2004), which we have adapted for use with prospective teachers.

Shari L. Stockero is an assistant professor of mathematics education at Michigan Technological University. Her research focuses on how to best support teacher learning and develop reflective practitioners, both in undergraduate teacher education courses and in professional development contexts.

Laura R. Van Zoest is a professor of mathematics education at Western Michigan University specializing in secondary school mathematics teacher education. She is particularly interested in the process of becoming an effective mathematics teacher and ways in which university coursework can best support that process.

Kazemi, E., Elliott, R., Lesseig, K., Mumme, J., Caroll, C. and Kelley-Petersen, M.
AMTE Monograph 6
Scholarly Practices and Inquiry in the Preparation of Mathematics Teachers
© 2009, pp. 171–186

11

Doing Mathematics in Professional Development to Build Specialized Content Knowledge for Teaching

Elham Kazemi
University of Washington

Rebekah Elliott
Kristin Lesseig
Oregon State University

Judy Mumme
Cathy Carroll
WestEd

Megan Kelley-Petersen
University of Washington

We share insights from our work in a project investigating what leaders learn about cultivating mathematically rich professional development environments for teachers. We discuss what leaders need to do when facilitating mathematical tasks with teachers if the goal is to develop teachers' specialized content knowledge (SCK) for teaching (Ball, Thames, & Phelps, 2008). SCK is the mathematical understanding and reasoning unique to teaching. With such a goal in mind, we argue for the importance of articulating the differences between doing mathematics in PD and in the K–12 classroom. Developing teachers' SCK for teaching requires that leaders design and use tasks in ways that

*orient teachers to SCK, foster particular
sociomathematical norms for explanation and
representational use, and orchestrate discussions to
make specialized knowledge explicit.*

Many professional development efforts in mathematics aim
to develop teachers' disciplinary knowledge because the quality
of teachers' mathematical knowledge plays a significant role in
the effectiveness of their instruction and the resulting student
learning (Hill, Rowen, & Ball, 2005). Yet what leaders[1] of
professional development (PD) actually need to know and be
able to do in their practice to support teacher learning of
mathematics is understudied and not well-defined, so much so
that Even's (2008) recent review of the literature on leader
practice focused on its "missing" literature.

In this paper we draw on our work in Researching
Mathematics Leader Learning (RMLL),[2] a five-year project
investigating what leaders learn about cultivating mathematically
rich PD environments for teachers. Specifically, we are
exploring what leaders need to learn to do when planning for and
facilitating mathematical tasks in PD in order for the activity to
be productive in developing teachers' mathematical knowledge.
When teachers do mathematics together in the PD setting, it is
easy for the focus of discussion to shift from purely
mathematical concerns to other pressing issues such as testing,
standards, instruction, student motivation, etc. These other
concerns can often shortchange the goal of developing
mathematical knowledge, depending on how the leader and
teachers pursue particular issues (Wilson & Berne, 1999).

In an effort to help leaders learn to create mathematically
rich discussions in the PD context, in RMLL seminars we
explored the construct of sociomathematical norms and a set of
practices for orchestrating productive mathematical discussions.
We thought that both of these frameworks, described more fully
below, would help leaders keep mathematical ideas central when
facilitating discussions in PD. Notably, both frameworks have
their origins in classroom research, and we adapted them for the
PD context. In addition, after completing one round of seminars

with leaders and observing a subset of them facilitate mathematical tasks with teachers, we have expanded our framework to include an articulation of the purposes for doing mathematics with teachers. The construct of specialized mathematical knowledge for teaching has now become central in our work, and we believe it will provide a sharpened focus for doing mathematics that is beneficial for professional educators.

In this article we offer insights we have gained by working with leaders and studying their attempts to make use of both sociomathematical norms and a set of practices for orchestrating productive mathematical discussions during their own facilitation of PD. These insights include: (a) the importance of articulating the differences between doing mathematics in PD and doing mathematics with students in the K–12 classroom, a difference that has caused us to think more carefully about our attempts to adapt frameworks developed from classroom research to PD; and (b) a need to link sociomathematical norms for explanation and the practices for orchestrating discussions to the purposeful development of specialized knowledge of mathematics for teaching (Ball, Thames, & Phelps, 2008).

Project Context and Conceptual Frameworks for Leadership Practice

We began with the hypothesis that paying explicit attention to the construct of sociomathematical norms would help leaders support the negotiation of mathematical reasoning and the development of teachers' mathematical knowledge as they conducted their own PD sessions. Leaders participated in a series of six full day seminars across the academic year. Each seminar began with solving and discussing a mathematical task that involved generalizing from arithmetic solutions as a way to investigate algebraic reasoning. Leaders' collective mathematical work and discussions of their methods for solving the task set the stage for the centerpiece of the seminar—a videocase of a mathematics PD leader engaging teachers in the same task. Tasks and videocases were drawn from a larger set of materials developed for leaders of mathematics PD (Carroll & Mumme,

2007). Leaders discussed both *what* mathematical explanations were shared in the videocase and *how* teachers in the videocase engaged in sharing explanations. We wanted leaders to notice how sociomathematical norms for explanation supported teachers' mathematical reasoning in the videocases. Between our seminars, leaders carried out PD with classroom teachers as part of their various leadership positions.

We created a frame for thinking about ways leaders could support teachers' mathematical understandings. With one part of this frame we aimed to help leaders distinguish social norms from sociomathematical norms (Yackel & Cobb, 1996). We designed seminar activities for leaders to pay attention to the nature of questioning, the treatment of errors and confusion, the locus and distribution of kinds of mathematical talk, and what mathematical work was accomplished by pursuing particular explanations (Kazemi & Stipek, 2001). We asked leaders to analyze PD environments for the social norms evident in whole group discussions as well as sociomathematical norms for explanation. For example, a productive social norm might be that teachers must describe and give reasons for their thinking, whereas a productive sociomathematical norm might be that an explanation must include a mathematical argument. Our focus led us to unpack four relevant aspects of sociomathematical norms for explanation: (1) sharing to emphasize the meaning of mathematical ideas and mathematical connections among solutions or representations, (2) articulating justifications for why and how particular methods work, (3) responding to confusion and errors by comparing ideas, re-conceptualizing problems, exploring contradictions, or pursuing alternative ideas, and (4) asking questions to deepen mathematical ideas.

A second part of the frame was a set of practices designed to help leaders think more holistically about orchestrating productive mathematical discussions. For this, we again drew on classroom-based research, specifically the work of Stein and colleagues (Stein, Engle, Smith, & Hughes, 2008). Based on their study of the implementation of cognitively demanding mathematical tasks, Stein et al. identified five practices teachers can employ as they more intentionally plan for and orchestrate discussions with students. With a similar goal of increasing leaders' intentionality, we adapted these practices to PD to engage leaders in: 1) *anticipating* teacher responses to rich mathematical tasks, 2) *monitoring* teachers' responses to the tasks during the explore phase, 3) purposefully *selecting* teacher work to share in whole group discussions, 4) purposefully *sequencing* the teacher work to be discussed, and 5) helping the group make mathematical *connections* between different teacher responses in order to develop powerful mathematical ideas. We refer to these throughout the rest of the paper as *practices for teacher sharing*.

The construct of sociomathematical norms and the practices for teacher sharing were included as a part of our initial seminar design for leaders. A third aspect of framing the work of doing mathematics emerged as we examined what occurred *during* our seminars and as we followed a subset of leaders into their own facilitation practice. During our data analysis, we recognized a need to identify more nuanced and detailed purposes for doing mathematics in PD and to discuss these purposes with leaders.

Researchers at the University of Michigan (Ball, Thames, & Phelps, 2008; see also http://sitemaker.umich.edu/lmt/home) have built a framework for understanding the mathematical knowledge entailed in teachers' work in the classroom that we find useful in our work with leaders. They have identified two kinds of mathematical knowledge that teachers need: Common Content Knowledge (CCK), the mathematical knowledge and skills used in professional settings other than teaching, and Specialized Content Knowledge (SCK), mathematical knowledge and skills uniquely needed by teachers. As we examine how leaders facilitate mathematical tasks in PD, we are specifically interested in instances where leaders try to build teachers' SCK. Teachers use and develop CCK when they solve mathematics tasks. They use and develop SCK when they generate and analyze the varied ways a particular task can be solved, the reasoning used to arrive at correct and incorrect solutions, mathematical connections between solutions, and the use of representations to convey mathematical meanings. To develop SCK teachers often need to unpack their mathematical understandings to develop a more flexible understanding of the ideas central to the school curriculum. For example, teachers, like other adults, can compute an answer to a multiplication problem, which requires common content knowledge. To develop specialized knowledge of multiplication, they need to unpack the various meanings of multiplication (e.g., grouping, skip counting, area), link written symbols to representations for both whole numbers and fractions (such as arrays), analyze common errors, and compare various solution methods (e.g., partial products, estimating and compensating). SCK is important for enhancing teachers' classroom instruction because they need to draw on it as they help students learn in the classroom (see Ball, Thames, & Phelps, 2008 for an elaboration).

Project Participants and Data Sources

Data for this report come from case studies from a subset of the 36 leaders who participated in RMLL seminars in two geographically distinct sites, which we refer to as Northwest (NW) and Southwest (SW). All leaders were volunteers from pre-existing groups charged with leading mathematics PD in various contexts with teachers, and most also worked with K–12 students in some capacity during the day. At the NW site two-thirds of the leaders had fewer than four years of facilitation experience and one third had over four years of experience as PD leaders. Two-thirds of these leaders worked with teachers across the K–12 spectrum. At the SW site the leaders had very little experience facilitating mathematics PD and worked mostly with elementary school teachers.

For all 36 leaders we documented engagement during our seminars by collecting their seminar work, transcribing videotapes of each session and distributing questionnaires before and after the seminars. We embedded the same task at the beginning and end of the seminar to compare how their ideas about facilitation had changed over the course of the seminar. We then followed a subset of leaders as they led their own professional development. We selected leaders who represented a range of grade levels, experience facilitating, and comfort with mathematics. The examples we use in this paper to illustrate our key insights come from case studies of four leaders, two from each participating site, who represent common experiences leaders had during the seminar and in their own facilitation of PD.[3]

Insights We Have Gained from Working with Leaders

Doing Mathematics in PD is Different from Doing Mathematics in the K–12 Classroom

Because there is so much overlap in the way the field thinks about doing mathematics in PD and doing mathematics in the classroom, including our own efforts to leverage classroom-based research for PD purposes, we think it is important to attend to what distinguishes doing mathematics in PD from doing mathematics in the K–12 classroom. These differences have direct bearing on what leaders do and what they hope to accomplish.

Teachers hold mathematical knowledge differently than students. Teachers often already "know" the mathematics. For the most part, they have the common content knowledge needed to solve the problems they give to their students and may have progressed at least once through the content as students and potentially many more times as teachers. From these experiences many teachers have developed compressed understandings that allow them to navigate a task, often jumping quickly to symbolic manipulations or algorithms (Adler & Davis, 2006) and yet potentially have missed opportunities to explore underlying reasoning (Lo, Grant, & Flowers, 2008). This became evident to us when listening to teachers (and leaders) complete tasks. For example, in a PD session facilitated by a leader, we heard one teacher offer a formula for finding the sum of consecutive integers saying, "I just learned way back when that if you want to sum them up you just take the next number and multiply the last number and the next number together and divide by 2. But I never visualized why that was working. I just knew it, so I wrote it down" (SW Case PD2, 2007). Many times teachers offered something they "remembered" to advance a group's progress on a problem. These offerings showed us how teachers draw on various experiences doing mathematics–as students, as teachers, and as learners in PD–to make sense of mathematics tasks. At the same time, teachers may recognize gaps in their knowledge and understanding, reflecting, as the teacher did above, that their ideas are limited and partial.

Because teachers hold mathematical knowledge differently than students, leaders must (a) recognize how teachers show their varying, and sometimes incomplete, mathematical understandings, (b) find ways to attend to that range as they monitor teacher engagement and manage discussions, and (c) pose tasks in particular ways to explicitly unpack the big mathematical ideas crucial to teaching a particular topic. We saw a leader attempting to do this when setting up a problem by saying,

> I'm going to push you a little bit to try to come up with as many solutions as you can to solve this problem in the time you're given. But there's a little hitch here that if you come up with a nice algebraic expression, that's great. But I'm going to ask you for some models and some ways to document why and how it works. (NW PD2, 2007)

This statement on its own could be heard in a K–12 classroom, but what gives it import in the PD context is that it is designed to begin the process of cataloguing and making sense of a set of approaches to a problem, something teachers do as they develop SCK. A student does not necessarily need to have command of the full range of solutions and their meanings. We propose that it may be important for leaders to explicitly articulate why teachers are being asked to generate a variety of models and how they relate back to symbolic expressions as a way to overtly negotiate sociomathematical norms for explaining with representations. As teachers make instructional decisions in the classrooms, this specialized content knowledge can help them with a range of the central tasks of mathematics teaching such as responding to students' questions, understanding the logic behind errors, and selecting representations for particular purposes.

   Leaders' relationships with teachers are different from teachers with students. Leaders' relationships with teachers in PD have an added layer of complexity. When doing mathematics together, some teachers may feel like they should or already do know the mathematics and may resist completing mathematical tasks or try to cover up any uncertainty or confusion. One leader explained, "I was torn as to whether or not to speak up, but I didn't want to put them on the spot if they weren't comfortable sharing out about their struggle through that" (SW Case Leader Interview, 2007). Leaders and teachers can narrate themselves and each other into particular kinds of identities with respect to doing mathematics, and leaders need to be aware of and attend to the ways that teachers position themselves and use status differentials during PD. Teachers often mark their contributions to discussions in one or more of the following ways: (a) I'm not a math person; (b) Those teachers know math because they are the middle school teachers (and we're elementary teachers); (c) That's Michael talking—he's really smart (i.e., in some PD contexts men can take and hold more status); and (d) I don't know math but I'm learning, and here's a question that occurs to me.

   In some cases the status differentials open up opportunities for learning. For example, when someone says, "I'm not good at math" and asks what she/ he believes to be a naïve question, this has the potential to engage everyone in deep analysis of the mathematics and to develop the specialized knowledge teachers need. Other times, it closes down opportunities, such as when elementary teachers defer to middle school teachers or when a secondary teacher dominates conversations and makes the mathematical questions seem obvious and self-explanatory. Leaders may be able to disrupt these displays of status by explicitly communicating to teachers that the work they are doing with mathematics is not just about finding solutions but about developing a kind of mathematical knowledge that they need as professionals that they did not gain from being students. As a field, we do not yet have a way of noting that we are working on developing specialized knowledge. Therefore, teachers' motivation for engaging in mathematics during PD may not fit squarely with this view of needing to develop SCK. Leaders may not necessarily hold this view either.

Orchestrating Productive Mathematical Discussions in PD

Leaders and teachers' mathematical work traverses a complex terrain, requiring leaders to have clear understandings of specialized knowledge for teaching so they can support teachers in developing this knowledge (Adler & Davis, 2006). Leaders and teachers must negotiate teachers' previous understandings of mathematics, yet still interactively co-construct productive sociomathematical norms that support teachers in investigating mathematics in ways that deepen specialized mathematical knowledge. To do this, leaders need to be able to unpack their own understandings of mathematics as well as engage teachers in doing so. They need to be able to anticipate what teachers might do with a task that would convey their common content knowledge and be ready to identify and attend to aspects of the mathematical ideas that need to be worked on explicitly. We recognize that leaders may not see unpacking teachers' understandings of mathematics as an explicit purpose for doing mathematics in PD. Thus, our work has led us to consider how these purposes get articulated in PD in ways that help teachers see the relationship between doing the mathematics and their work with students and how tasks should be framed with teachers when doing mathematics in PD. This more focused way of thinking about purposes for doing mathematics in PD provides direction to the way we might engage leaders in making sense of and fostering particular sociomathematical norms and using practices for teacher sharing. In particular, a more purposeful focus on developing SCK has significant implications for how leaders use tasks and questioning to elicit particular kinds of unpacked explanations. The goal of developing SCK can also sharpen the kinds of mathematical connections made while solutions are shared and discussed.[4] In the following section we share insights we have gained from analyzing our leaders' facilitation.

Selecting aspects of mathematics to pursue. We noticed by watching leaders facilitate PD sessions that the process of identifying the big mathematical goal(s) can be challenging. Leaders try to balance what they understand to be needs of their teachers, their own mathematical understanding, and how to use the time they have. They also wrestle with where to place the focus in discussing tasks with teachers. One leader identified several major mathematical ideas that were at play in the task he posed: the move from particular arithmetical computations to generalized ideas about operations, the meaning and use of variable in symbolic expressions, and the meaning of the distributive property. He was trying to unpack the mathematical ideas through the way he sequenced solutions so that teachers could see how computation methods could be generalized and how all solutions tied to a model of the situation. Through discussion across the teachers' solutions, he wanted the teachers to realize that assigning variables to represent different parts of the model could result in different, yet equivalent, expressions. He also noticed that he could raise issues about the meaning of the distributive property as teachers shared the equivalence of various algebraic expressions they had generated for the task. This is an impressive list and perhaps more appropriate in developing a trajectory of mathematical ideas to work on over several sessions rather than one, something that the leader also noted in his reflections on the session.

Doing more than improving teachers' individual and social orientations toward mathematics. We found the identification of goals for doing mathematics was mediated by leaders' attention to cultivating productive individual and social orientations that guide teachers' participation in mathematical activity. As a result, one group of leaders saw the development of particular orientations as a primary goal of doing mathematics in PD. For example, in one PD session we observed, the leaders acted on an overriding priority within their district to help teachers understand the constructivist learning philosophy and how it undergirded their new curriculum adoption. Doing mathematics in PD was thus situated within this major aim. One leader explained:

> What we were looking at with the staircase problem was
> to engage them in that struggle as well as to provide
> some modeling about work within the constructivist
> model. It's not about going up to someone and saying,
> why don't you try this, you know when we were
> interacting with the groups, but it was about that
> questioning, and that whole engagement process. (SW
> Leader Interview, 2007)

The leaders wanted the mathematical task to generate
willingness to persevere with problem solving in the face of
difficulty, to be comfortable sharing vulnerabilities, and to
cultivate the curiosity to question each other and engage in the
task. A particular subject matter goal was missing. One of our
challenges as we work with PD leaders is how to help them
cultivate these orientations with teachers while also remaining
focused on specific mathematical goals when discussing
solutions to tasks.

Thinking about tasks and task use over time. As we have
watched leaders facilitate PD we have begun to consider how we
might help leaders anticipate ways to press for particular kinds of
explanations based on a clear identification of the kinds of SCK
they aim to develop when working on a specific mathematical
task. Our initial observations showed us that we needed to be
more carefully attuned to helping leaders think about the tasks
they choose and how they use them within their time constraints
to build SCK. We have begun to consider how we help leaders
plan for the kinds of conversations they would like to see
happen, adequately attend to the time needed for these kinds of
conversations, and help them consider the mathematical
trajectory across several PD sessions.

One major insight we have gained thus far is how task
choice and discussion prompts are intimately related to the kinds
of knowledge to be developed. Some of our leaders previously
had used mathematical tasks from the adopted school curriculum
in PD. They could assume teachers had ways of solving the
problems because they had used them with students. However, to
develop SCK, teachers need to map the range of possible

solutions and errors for a task and the reasoning they used to obtain those solutions and errors rather than finding only one solution to each task. Developing skilled use of representations is also part of SCK. For example, teachers could develop knowledge of how the number line can be used to represent both distance and removal models for subtraction by working with particular subtraction problems and considering the mathematical meanings of those models in different situations.

Leaders from our elementary-dominated site were understandably concerned with teacher engagement and whether teachers appreciated the struggle that learners undergo when learning new content. To help teachers appreciate this struggle, the leaders chose a mathematical task they had done during one of our seminars, one more typical at the late middle or early high school level that was relatively challenging for the group of elementary school teachers. The choice of this task *did* allow teachers to experience struggle and appreciate the role of struggle in constructing knowledge (and, in their case, in understanding something about constructivism as a learning philosophy). But this activity may not have helped teachers develop SCK for mathematical domains that are critical to the elementary school curriculum. The leaders were also very concerned with teacher engagement with tasks. For example, if the task is perceived as too simple, what does a leader do to help teachers dig into the task? Based on our understanding of the specialized mathematical work entailed in teaching, we are experimenting with more suitable prompts to invoke and develop SCK that may include asking teachers to: (a) examine the reasoning behind solutions rather than just solve the problems themselves; (b) understand the logic behind errors; or (c) show how varied solutions can be represented with particular models or tools (Ball, Thames, & Phelps, 2008).

Conclusion

Working with PD leaders has led us into uncharted territory. One of the most consequential insights we have gained is the need to marry ideas about clear purposes for doing mathematics with the kind of knowledge we want teachers to gain and the practices we use in PD. We need to learn how to frame the purpose(s) for doing mathematics in PD better. The distinction between common content knowledge and specialized content knowledge has been generative in helping us think about revising our seminars to better support leaders. In particular, we hope to help leaders relate the mathematical tasks and prompts they use in PD to the particular kinds of explanations and justifications they hope to elicit from teachers and how those unpacked explanations are useful for teachers' work in classrooms. With more clarity in purpose, we see the potential for leaders to facilitate mathematical tasks so that teachers' work remains sharply focused on building teachers' specialized mathematical knowledge.

References

Adler, J., & Davis, Z. (2006). Opening another black box: Researching mathematics for teaching in mathematics teacher education. *Journal for Research in Mathematics Education, 37*, 270–296.

Ball, D. L., Thames, M. H., & Phelps, G. (2008). Content knowledge for teaching: What makes it special? *Journal of Teacher Education, 59,* 389–407.

Carroll, C., & Mumme, J. (2007). *Learning to lead mathematics professional development.* Thousand Oaks, CA: Corwin.

Even, R. (2008). Facing the challenge of educating educators to work with practicing mathematics teachers. In T. Wood, B. Jaworski, K. Krainer, P. Sullivan, & T. Tirosh (Eds.), *The international handbook of mathematics teacher education: The mathematics teacher educator as a developing professional* (Vol. 4; pp. 57–74). Rotterdam, The Netherlands: Sense.

Hill, H., Rowen, B., & Ball, D. L. (2005). Effects of teachers' mathematical knowledge for teaching on student achievement. *American Educational Research Journal, 20*(2), 371–406.

Kazemi, E. & Stipek, D. (2001). Promoting conceptual thinking in four upper-elementary mathematics classrooms. *The Elementary School Journal, 102*(1), 59–80.

Lo, J., Grant, T. J. & Flowers, J. (2008). Challenges in deepening prospective teachers' understandings of multiplication through justification. *Journal of Mathematics Teacher Education, 11*(1), 5–22.

Stein, M. K., Engle, R. A., Smith, M. S., & Hughes, E. K. (2008). Orchestrating productive mathematical discussions: Five practices for helping teachers move beyond show and tell. *Mathematical Thinking and Learning, 10*, 314–340.

Wilson, S. M., & Berne, J. (1999). Teacher learning and the acquisition of professional knowledge: An examination of research on contemporary professional development. In A. Iran-Nejad & P. D. Pearson (Eds.), *Review of Research In Education* (pp 173–209). Washington, DC: American Educational Research Association.

Yackel, E. & Cobb, P. (1996). Sociomathematical norms, argumentation, and autonomy in mathematics. *Journal for Research in Mathematics Education, 27*, 458–477.

Endnotes

[1] We use the term "leader" to refer to the person who facilitates mathematics professional development.

[2] The research is supported by a grant from the National Science Foundation (NSF) (ESI-0554186). Opinions expressed in this report are the authors' and do not necessarily reflect the views of NSF.

[3] A representative sample of leaders was selected from the larger group.

[4] We also recognize how our work in doing mathematics in PD and developing SCK then connects to how SCK is deployed in service of learning other domains of knowledge needed for teaching – such as knowledge of content and curriculum, content and students and content and teaching. The depth of teachers' SCK would seem to impact how they engage in other

common PD tasks such as examining student work, planning lessons, evaluating teacher moves in discussions, etc.

Cathy Carroll is a senior research associate at WestEd. Her work involves designing and facilitating leadership development programs for K–12 mathematics leaders.

Rebekah Elliott is an assistant professor at Oregon State University. Her teaching and research centers on learning to facilitate mathematics for learners in K–12 schools, professional education, and free choice environments.

Elham Kazemi is an associate professor of mathematics education at the University of Washington. Her research interests include studying teacher learning and designing professional education at all levels to improve classroom practices.

Megan Kelley-Petersen is a doctoral student at the University of Washington. Her research interests are focused on supporting teachers to develop ambitious instructional practices in elementary mathematics.

Kristin Lesseig is a doctoral student at Oregon State University. Driven by classroom teaching experiences, her interest is in providing opportunities for both preservice and inservice teachers to develop mathematical understandings necessary for teaching.

Judy Mumme is a program director at WestEd. Her work involves designing and developing mathematics leadership experiences for K–12 leaders as well as researching leader learning.

Lager, C. A.
AMTE Monograph 6
Scholarly Practices and Inquiry in the Preparation of Mathematics Teachers
© 2009, pp. 187–201

12

Huerto de Manzanas: Reading Comprehension Awareness for Secondary Mathematics Teachers of English Learners

Carl A. Lager
University of California, Santa Barbara

Though understanding a mathematics problem is the first step towards solving it (Pólya, 1945)*, many secondary English learners (ELs) wrestle with making meaning of written word/story problems on their own. Because many homework problems and most assessment tasks are done individually, developing a learner's independent reading comprehension proficiency for mathematics problem solving is critical. Though the National Council of Teachers of Mathematics* (NCTM, 2000) *expects secondary mathematics teachers to help students learn to read text carefully, almost no practice-based guidance exists in our field to help teachers do so. To raise awareness and spur creation of professional development, the Huerto de Manzanas intervention was created, facilitated, and researched. The intervention structure and research findings are shared.*

Huerto de Manzanas

Preparing pre-service secondary mathematics teachers and continuing to support the professional development of their in-service colleagues are necessary and worthwhile investments for the mathematics education of all students (Darling-Hammond & Bransford, 2005; Hill & Ball, 2004), but especially our English learners (ELs). ELs are learners with a first language other than English who are in the process of acquiring grade level English

(Educational Alliance, 2008). They are receiving greater attention from our field than ever before, in part, because of their explicit inclusion in current state accountability systems under No Child Left Behind (NCLB, 2002) and the persistent mathematics achievement gaps on large-scale assessments when compared to non-ELs (National Assessment of Educational Progress [NAEP], 2008). In addition, their numbers continue to grow, with an estimate of 2.1 million EL students in grades 6–12 (Kindler, 2002; US Department of Education [USDOE], 2004).

Though some secondary EL mathematics learning needs have been identified and studied over the past 20 years (e.g., Morales, 2004; Moschkovich, 2002; Spanos, Rhodes, Dale, & Crandall, 1988), reading comprehension has received scant attention in comparison (Celedon-Pattichis, 2003). Though understanding a mathematics problem is the first step towards solving it (Pólya, 1945), many ELs wrestle with making meaning of written word/story problems on their own. Though the National Council of Teachers of Mathematics (NCTM, 2000) expects secondary mathematics teachers to help students learn to read text, the most current related work for secondary ELs in mathematics classrooms does not focus on individual student-text interactions (e.g., Fischer & Perez, 2008; Coggins, Kravin, Coates, & Carroll, 2007). Therefore, to help secondary mathematics teachers understand the strengths and needs of students from diverse linguistic backgrounds and respond to NCTM's (2000) expectation, the Huerto de Manzanas (HdM) intervention was created to raise awareness of secondary EL reading challenges. During the intervention, participants engage with modified Spanish versions of a large-scale mathematics item written originally in English. Lessons learned are shared during the small and whole group discussions that immediately follow.

Framework for Developing the Intervention

Loucks-Horsley, Love, Stiles, Mundry, and Hewson's (2003) professional development design framework was used to iteratively develop, facilitate, analyze, and revise the HdM intervention. The framework's six steps—Commit to vision and standards, Analyze student learning and other data, Set goals, Plan, Do, and Evaluate—are fleshed out below.

Committing to Vision and Standards

The vision was a 45 minute professional development intervention that raises participant awareness of the reading challenges secondary ELs typically encounter when engaging individually with written mathematics tasks. The intervention's focal mathematics task is a released item from the California High School Exit Exam (CAHSEE, 2007). The item is written for California Mathematics Standard 7AF4.1 (Curriculum Development and Supplemental Materials Commission – Mathematics [CDSMCm], 2006)[1] but is also aligned with two of NCTM's (2000) Grades 6–8 Algebra Expectations[2]. For adult learning, the intervention's content and structure explicitly draw from NCTM's (2007) recommendations for the education and continued professional growth of teachers of mathematics. Standard 3, Knowledge of Students as Learners of Mathematics, includes the role language plays in student understanding, especially for ELs. Standard 5, Participation in Career-Long Professional Growth, calls for professional development tasks to be constructed around artifacts (in this case, the released item) so that "teachers are led to analyze, raise questions, experience disequilibrium, and progress in their practice" (160).

Analyzing Student Learning and Other Data

EL mathematics assessment performance data and research were explicitly incorporated into the intervention (e.g., NAEP, 2008). English and Spanish task-based reading comprehension-mathematics interactions with secondary ELs (Lager, 2006; Celedón-Pattichis, 2003) were considered when generating the Spanish, modified versions of the English mathematics item.

Setting Goals

Raising participant awareness of reading challenges secondary ELs of varying English language proficiencies typically encounter on written tasks was the immediate intervention goal. Using such awareness as a catalyst to pay explicit attention to reading comprehension-mathematics interactions to improve teaching practice and mathematics task development was the long-term goal.

Planning

To meet these professional learning goals, a seven-part workshop strategy was planned.

Part 1. A brief Powerpoint™ presentation (7–10 minutes) provides basic information to set the context. Showing EL demographics, EL/Non-EL achievement gaps, and the shift in the mathematics education community from an EL deficit framework (Schorling, 1926) to one of accommodation and inclusion (NCTM, 2000) are examples of shared data.

Part 2. Participants engage with four related versions of one released large scale high school exit exam item (CAHSEE, 2007). Spanish is chosen for versions 1–3 because the wide spread of Spanish proficiency among the participants typically mirrors the range of ELs' English proficiency in mathematics classrooms. Version 3 is the Spanish version of the actual English item (Version 4).

For each version, participants have 90 seconds to silently and independently read the item (projected on a screen), solve it, and write down the answer. At the conclusion of each 90-second interval, participants write down their level of confidence in their answer (1–5), with 1 meaning not confident and 5 meaning confident. Then, that version of the item is replaced with the next version and projected. In six minutes, this process repeats for all four versions of the item.

Each version of the item is listed in Figure 1. Readers can simulate the experience in the HdM workshop by using two sheets of paper to hide the text that precedes and follows a version of the problem so that only one version of the item can be seen at a time, allowing yourself 90 seconds to process each version.

V1 – El dueño de un huerto de manzanas manda sus manzanas en cajas. Cada caja vacía pesa k kilogramos (kg). El peso medio de una manzana es a kg y el peso total de una caja llena de manzanas es b kg. ¿Cuántas manzanas han sido empacadas en cada caja?

V2 – El dueño de un huerto de manzanas manda sus manzanas en cajas. Cada caja vacía pesa k kilogramos (kg). El peso medio de una manzana es a kg y el peso total de una caja llena de manzanas es b kg. ¿Cuántas manzanas han sido empacadas en cada caja?

   A) $b + k$ B) $(b - k) / a$ C) b / a D) $(b + k) / a$

V3 – El dueño de un huerto de manzanas manda sus manzanas en cajas. Cada caja vacía pesa 2 kilogramos (kg). El peso medio de una manzana es 0.25 kg y el peso total de una caja llena de manzanas es 12 kg. ¿Cuántas manzanas han sido empacadas en cada caja?

   A) 14   B) 40   C) 48   D) 56

V4 – The owner of an apple orchard ships apples in boxes that weigh 2 kilograms (kg) when empty. The average apple weighs 0.25 kg, and the total weight of a box filled with apples is 12 kg. How many apples are packed in each box?

   A) 14   B) 40   C) 48   D) 56

Figure 1. Four versions of Huerto de Manzanas item.

Part 3. Participants get four minutes to freewrite individually to the following questions: What meaning-making and problem solving strategies did you use? How effective were they? What "mental movies" did you generate? How did you feel? How confident were you?

Parts 4, 5, 6, and 7. Participants get 5–7 minutes to share insights and feelings with their neighbors/tablemates (Part 4).

During that time, the facilitator walks around to listen to the points shared at different tables to determine how to best orchestrate the subsequent 8–10 minute whole group discussion (Part 5). Participants then put their teachers' hats on, look at Version 4, and in small groups consider comprehension challenges ELs would likely encounter (Part 6). After five minutes, participants share their predictions (Part 7).

Doing and Evaluating

Facilitated 11 times to a total of more than 500 participants, the HdM (translated as *Apple Orchard*) intervention was continually evaluated, and revised throughout a 12-month period. Within Guskey's (2000) five-level professional development evaluation model, only Levels 1 and 2 apply to this intervention. How participants considered their HdM experiences is Level 1. Level 2 is how and to what extent participants attained the intervention's intended affective and cognitive learning goals.

Applying Blau's (2003) data gathering and self-evaluation techniques for professional development workshops, the HdM intervention was similarly analyzed. The data sources included the slides and handouts from the interventions themselves (see www.huertodemanzanas.com), videotaped versions of the intervention, evaluations from the participants, personal communications with participants, and facilitators' memory. Key findings summarize how participants engage with Versions 1–3 of the item, their reading comprehension strategies, and how and to what extent Level 1 and Level 2 goals were met.

Findings

Version 1

Version 1 is a closed, constructed response item, where there exists a single path to a single solution that must be generated solely from the information given (Romagnano, 2006). Yet, most participants do not recognize its type because of the intentionally created cognitive challenges: no answer options, all the quantities are variables, written in Spanish, and no accompanying visual representations. However, many participants did recognize some or all of the variables, leading to the generation of incorrect answers like $a + b$ and $(b - 2a) / k$.

Though most attempt to wrestle with Version 1, 10–20% of participants break the rules of engagement. Looking at a neighbor's work/answer or initiating a conversation with a tablemate are typical reaching out actions. The rule breakers usually feel confident about their own mathematical competencies but not their language proficiencies. Therefore, they seek out assistance, often from persons they believe are Spanish proficient. However, a few participants with little or no Spanish proficiency resist silently in English (e.g., reading a newspaper, texting a friend).

Rule breakers and resistors want to solve the mathematics problem, but feel they cannot sufficiently access the item by themselves. Across the board, participants usually report their lowest levels of answer confidence at this point. Wanting to problem solve, feeling isolated, lacking confidence, and wanting to resist are intended Level 2 affective states experienced by participants. Their feelings and actions mirror those of many ELs in mathematics classrooms.

Version 2

Participants recognize Version 2 is a closed, selected-response item (Romagnano, 2006), meaning there's a single path to a single solution but the solver can select the correct answer from a list of possible responses. The purposeful inclusion of the four answer choices led participants to EL-like cognitive experiences and affective states.

Participants who didn't understand Version 1 often used the four answer choices to work backwards toward a solution; their confidence usually rose. Those who thought they had understood Version 1 said seeing their Version 1 answer among the given answer choices increased their confidence. Those who did not see their Version 1 answer among the four answer choices lost confidence, but recognized they had meaning-making and/or problem solving errors to correct. Worst off were the "false positive" participants who saw their initially incorrect answer to

Version 1 show up as a distracter in Version 2; their confidence wrongfully rose.

Version 3

Participants see that Version 3 is the same as Version 2, except the variables have been replaced with numbers. Though the focus shifts from the relationship among the problem's quantities to manipulating the quantities themselves, many struggling participants continue to work backwards toward determining a solution. Despite having seen three versions of the same item, many participants with beginning to intermediate Spanish proficiency continue to report low to medium confidence levels. They report being frustrated because they feel mentally exhausted, rushed due to time pressure, and/or less intelligent than their Spanish-proficient colleagues. These Level 2 affective states experienced by the participants are intended.

Reading Comprehension Strategies across Versions

Because many participants were frustrated when reading the Spanish versions of the item, like ELs often do in English, they use different reading comprehension strategies to access the item. Based on prior experience, participants' usually focus on identifying and understanding information they believe relevant to solving the problem. With sentences as a unit of analysis, participants often look to the last sentence first (the question) to determine what to solve. However, after struggling with Version 1, many participants look at the answer choices first in Versions 2 and 3 to try to reverse engineer the question. Conversely, because they know the first sentence usually sets the context but often is not crucial to answering the item's question, some participants invest very little, if any, time understanding the first sentence.

Focusing on words, most participants recognize that knowing every word's meaning is unnecessary to understand and solve the problem. Still, across Versions 1–3, many participants look for cognates (*kilogramos*/ kilograms), recurring words (*manzana, caja, peso*/apple, box, and weight*)*, and words they think they know (*peso, media, y*) to create cognitive footholds into the problem. Though on the surface such strategies are sound, pitfalls can lie underneath.

In this context, some participants think *peso* means the Mexican unit of money (instead of weight) and *media* means half (instead of average). Similarly, though focusing on the meaning of the item's variables (a, k, b) helps many understand Versions 1 and 2, a few participants mistakenly consider y a variable or mistranslate it as "+," instead of the conjunction "and." As a result, participants' incomplete understandings become further muddled. They generate logical but incorrect conceptions of the problem, such as visualizing apples being sliced in two or solving for the cost per apple or box.

In fact, just one unknown word or phrase can detour a solver. For example, not knowing *vacía* means "empty" leads some participants to not subtract the weight of the empty box from the weight of the full box. Yet, sometimes a word's form is more important than its meaning. For example, the *manzanas* themselves aren't as important as how many there are, how they are acted upon, and how they are measured. Instead of figuring out *manzanas* means apples, a few participants substitute familiar objects, like bananas or marbles, and proceed forward.

Level 1 – How Participants Considered HdM

Considering their experiences as a whole, approximately 70–80% of the participants rated the workshop highly. Though it is not known to what extent each participant's awareness was heightened, many commented that they had temporarily felt like ELs and wanted to know what to do to better meet their ELs' reading comprehension needs and to write better mathematics assessment items in English. Therefore, for them, the Level 1 objectives were met.

However, the other 20–30% did not rate it highly. They usually said that they did not see the point, that the activities were too drawn out, or that such language teaching was not their responsibility. Though there is a risk that this participant subset may be less sensitive to the needs of ELs in the future, further entrenchment of their "not-my-problem" belief system is more likely. These results are not surprising because this activity tacitly pushes on several overarching constructs that are personal, contentious, and intertwined: language, culture, power, race, xenophobia, assimilationism, and enculturationism. Addressing these constructs requires professional development that goes far beyond the scope and duration of the HdM intervention.

Level 2 – HdM's Affective and Cognitive Goals

Most participants temporarily experience some of the emotions ELs regularly experience when they engage with mathematics items in English, such as being frustrated and feeling pressured. Upon reflection, they consider how their feelings in a Spanish language assessment environment prompt their own problem solving, rule breaking, and resistance behaviors. Experiencing these brief emotion-action connections helps participants better understand similar EL motivations and actions. Therefore, the Level 2 affective goals were met.

In addition, participants achieve four Level 2 cognitive goals. First, they raise their awareness of their Spanish reading comprehension challenges and the meaning making strategies they used to address those challenges. Second, participants start considering what they do and do not know about the English reading challenges encountered by ELs engaging individually with written mathematics tasks. For example, participants recognize that sometimes ELs get stuck on an unknown word or phrase they don't recognize as being immaterial to solving the problem. When predicting potential EL challenges with the English version of the HdM item, participants regularly identify *owner* and *orchard* as those kinds of words. However, few see the uncommon phrase (*average apple*), the noun-verb shift (*ships* – meaning to send out, not boats on the water), and the passive voice construction (*are packed*) as confusing for some ELs.

Third, most participants realize they do see ELs use reading comprehension strategies to engage with written mathematics problems. Discussing the task with a classmate (when allowed), looking for recognizable information within the task, and working backwards from provided answer choices are three common ones. In terms of their own teaching, participants report that they often define mathematical terms and teach "*is* means =" translation strategies. However, participants do recognize ELs require greater metalinguistic support than they currently provide. Reaching this fourth cognitive goal prompts participants to consider ways to help ELs access grade level tasks. Helping ELs independently identify, connect, and manipulate relevant information and filter out irrelevant information are suggested supports.

Conclusion

Because most secondary mathematics teachers are native English speakers who learned mathematics in English, many are unaware of the type, number, and depth of reading comprehension challenges ELs encounter with written mathematics tasks. Yet, to accommodate English proficiency differences among students effectively and sensitively, teachers need to understand and confront their own beliefs and biases (NCTM, 2000). Therefore, the Huerto de Manzanas intervention was created to be a first step toward transforming participants' practice (Mezirow, 1997). Next steps should help secondary mathematics teachers move from brainstorming potential EL reading comprehension challenges to explicitly identifying and addressing them (Celedón-Pattichis, 2004). Helping mathematics teachers teach ELs to use reading comprehension strategies to strengthen their meaning making skills (e.g., Lager, 2008; Thompson, Kersaint, Richards, Hunsader, & Rubenstein, 2008; Schleppegrell, 2007) can help close EL/Non-EL opportunity and achievement gaps (Flores, 2008). Telling students to read a problem carefully without teaching them how is unfair and counterproductive.

Author Note: I thank Megan Staples, Bill Jacob, and my reviewers for their insightful feedback on this work.

198

Lager

References

Blau, S. (2003). *The literature workshop: Teaching texts and their readers*. Portsmouth, NH: Heinemann.

California Department of Education (CDE). (2008). *Number of English learners by language*. Retrieved May 16, 2008 from http://dq.cde.ca.gov/dataquest/ LEPbyLang1.asp?cChoice= epbyLang1&cYear=2006-07&cLevel=State&cTopic= C&myTimeFrame= S&submit1=Submit.

California High School Exit Exam (CAHSEE). (2007). *Mathematics released test questions*. Sacramento, CA: California Department of Education. Retrieved May 31, 2008 from http://www.cde.ca.gov/ta/tg/hs/documents/math07rtq.pdf

Celedón-Pattichis, S. (2004). Research findings involving English-language learners and implications for mathematics teachers. In M. F. Chappell, J. Choppin, & J. Salls (Eds.), *Empowering the beginning teacher of mathematics: High school* (p. 45–46). Reston, VA: National Council of Teachers of Mathematics.

Celedón-Pattichis, S. (2003). Constructing meaning: Think-aloud protocols of ELLs on English and Spanish word problems. *Educators for Urban Minorities*, 2(2), 74–90.

Coggins, D., Kravin, D., Coates, G. D., & Carroll, M. D. (2007). *English language learners in the mathematics classroom*. Thousand Oaks, CA: Corwin Press.

Curriculum Development and Supplemental Materials Commission–Mathematics (CDSMCm) (2006). *Mathematics framework for California public schools, kindergarten through grade twelve*. Sacramento, CA: California Department of Education.

Darling-Hammond, L., & Bransford, J. (2005). *Preparing teachers for a changing world: What teachers should learn and be able to do*. San Francisco, CA: Jossey- Bass.

Educational Alliance. (2008). Retrieved May 20, 2008 from http://www.alliance.brown.edu/tdl/ell.shtml

Fischer, J., & Perez, R. (2008). Understanding English through mathematics: A research based ELL approach to teaching all students. In R. Kitchen & E. Silver (Eds.), *Research Monograph of TODOS: Mathematics for All, 1,* 43–60. National Education Association.

Flores, A. (2008). The opportunity gap. In R. Kitchen & E. Silver (Eds.), *Research Monograph of TODOS: Mathematics for All, 1,* 1–18. National Education Association.

Guskey, T. R. (2000). *Evaluating professional development.* Thousand Oaks, CA: Corwin Press.

Hill, H. C., & Ball, D. L. (2004). Learning mathematics for teaching: Results from California's mathematics professional development institutes. *Journal for Research in Mathematics Education, 35*(5), 330–351.

Kindler, A. L. (2002). *Survey of the states' limited English proficient students and available educational programs and services 2000–2001 Summary Report.* Office of English Language Acquisition. Retrieved February 1, 2009 from: http://www.ncela.gwu.edu/policy/states/reports/seareports/0001/sea0001.pdf

Lager, C. A. (2006). Types of mathematics-reading interactions that unnecessarily hinder algebra learning and assessment. *Reading Psychology,* 27(2–3), 165–204.

Lager, C. A. (2008). Reading comprehension for algebra learning. *California Mathematics Council (CMC) ComMuniCator, 32(4),* 36–41.

Loucks-Horsley, S., Love, N., Stiles, K. E., Mundry, S., & Hewson, P. W. (2003). *Designing professional development for teachers of science and mathematics* (2nd ed.). Thousand Oaks, CA: Corwin Press.

Mezirow, J. (1997). Transformative learning: Theory to practice. In P. Cranton (Ed.), *Transformative learning in action: Insights from practice* (pp. 5–12). New Directions for Adult and Continuing Education No 74. San Francisco: Jossey Bass.

Morales, H. (2004). *A naturalistic study of mathematical meaning-making by Latino high school students.* Chicago: University of Illinois at Chicago.

Moschkovich, J. N. (2002). A situated and sociocultural perspective on bilingual mathematics learners. In N. Nassir

and P. Cobb (Eds.) *Mathematical thinking and learning.*
    Special Issue on Diversity, Equity, and Mathematical
    Learning, 4(2&3), 189–212.
National Assessment of Education Progress (NAEP). (2008).
    NAEP Data Explorer. Retrieved May 14, 2008 from:
    http://nces.ed.gov/nationsreportcard/nde/criteria.asp
National Council of Teachers of Mathematics. (NCTM) (2000).
    Principles and standards for school mathematics. Reston,
    VA: The National Council of Teachers of Mathematics.
National Council of Teachers of Mathematics. (NCTM) (2007).
    Standards for the education and continued professional
    growth of teachers of mathematics. In T. S. Martin (Ed.),
    Mathematics teaching today: Improving practice, improving
    student learning (pp. 109–170). Reston, VA: The National
    Council of Teachers of Mathematics.
No Child Left Behind Act of 2001. Pub. L. No. 107-110, 115
    Stat. 1425 (2002).
Pólya, G. (1945). *How to solve it.* Princeton, NJ: Princeton
    University Press.
Romagnano, L. (2006). *Mathematics assessment literacy:*
    Concepts and terms in large-scale assessment. Reston, VA:
    The National Council of Teachers of Mathematics.
Schleppegrell, M. J. (2007). The linguistic challenges of
    mathematics teaching and learning: A research review.
    Reading & Writing Quarterly, 23(2), 139–159.
Schorling, R. (1926). Suggestions for the solution of an
    important problem that has arisen in the last quarter of a
    century. In R. Schorling (Ed.), *A general survey of the*
    progress of mathematics in our high schools in the last
    twenty-five years: Yearbook 1 (pp. 58–105). Oak Park: The
    National Council of Teachers of Mathematics.
Spanos, G., Rhodes, N., Dale, T., & Crandall, J. (1988).
    Linguistic features of mathematical problem solving. In R.
    Cocking & J. Mestre (Eds.), *Linguistic and cultural*
    influences on learning mathematics (pp. 221–240). Hillsdale,
    NJ: Lawrence Erlbaum Associates.
Thompson, D. R., Kersaint, G., Richards, J. C., Hunsader, P. D.,
    & Rubenstein, R. N. (2008). *Mathematical literacy: Helping*

students make meaning in the middle grades. Portsmouth, NH: Heinemann.

United States Department of Education (USDOE). (2004). Fact Sheet: NCLB provisions ensure flexibility and accountability for Limited English Proficient students. Retrieved May 20, 2008 from: http://www.ed.gov/nclb/accountability/schools/factsheet-english.html

Endnotes

[1] Grade 7 – Algebra and Functions: Solve two-step linear equations and inequalities in one variable over the rational numbers, interpret the solution or solutions in the context from which they arose, and verify the reasonableness of the results.

[2] Grades 6–8 Algebra: Use symbolic algebra to represent situations and to solve problems, especially those that involve linear relationships; model and solve contextualized problems using various representations, such as graphs, tables, and equations.

Carl A. Lager works in the Gevirtz Graduate School of Education at the University of California, Santa Barbara where he researches and develops prospective, pre-service, and in-service secondary mathematics teachers and teacher educators, with special attention paid to meeting the instructional and assessment needs of English learners in grades 6–12.

Ghousseini, H.
AMTE Monograph 6
Scholarly Practices and Inquiry in the Preparation of Mathematics Teachers
© 2009, pp. 203–218

13

Designing Opportunities to Learn to Lead Classroom Mathematics Discussions in Pre-service Teacher Education: Focusing on Enactment

Hala Ghousseini
University of Michigan–Ann Arbor

In this essay, I describe how discourse routines can be used as intellectual and practical tools to prepare pre-service teachers to learn to lead classroom mathematics discussions. This work illustrates how teacher education programs can focus not only on pedagogies of investigation but also on enactment and help prospective teachers think systematically about the complexity of practice.

University-based teacher preparation, as it is currently typically structured, has minimal influence on pre-service teachers' practice (Clift & Brady, 2005). This situation is attributed to many reasons; among the most prominent ones are the problems of enactment and complexity (Hammerness et al., 2005). Traditionally, teacher education programs have mainly focused on developing pre-service teachers' commitments to certain visions of teaching and learning, leaving to field experiences the actual work of learning pedagogical skill (Grossman & McDonald, 2008). One problem that faces pre-service teacher education is not only preparing pre-service teachers to "think" like a teacher, but also to put what they know into action. Kennedy (1999) argued that teacher education programs need to focus on equipping novices with the behavioral enactments that can help them translate the ideas they learn in their initial teacher preparation to practice.

Another problem that is associated with the problem of enactment in teacher education is the complex nature of practice (Hammerness et al., 2005). The vision of teaching projected by reformers, which encourages students to develop their own ideas and teachers to respond judiciously to those ideas, requires a lot of spontaneous judgments that cannot be specified in advance. With little knowledge about appropriate enactments in practice, the complexity and uncertainty in teaching end up crippling pre-service teachers' ability to take reasonable courses of action, and consequently reinforce their reliance on the frames of reference they develop through their apprenticeship of observation (Kennedy, 1999). Teacher education programs can help pre-service teachers think systematically about the complexity of practice by providing them with the intellectual and practical resources for use in the classroom (Hammerness et al., 2005).

My aim in this chapter is to contribute to our knowledge of the kind of intellectual and practical resources that can support novice mathematics teachers in leading productive classroom mathematics discussions. I describe an instructional intervention, in the context of a secondary mathematics methods course, which focused on discourse routines as intellectual and practical tools that can enable pre-service teachers to lead a classroom mathematics discussion, to learn about it, and to manage some aspects of its complexity.

Design Considerations

The design of opportunities to learn to lead productive classroom mathematics discussions for pre-service teachers needs to be based on a number of considerations including the nature of productive classroom mathematics discussions; the nature of teachers' work in leading such discussions; and the nature of pedagogies of enactment in teacher preparation.

The Nature of Productive Classroom Mathematics Discussions

Discussing mathematics in the classroom is an approach to teaching that honors the nature of mathematics as a form of human knowledge in the making (Polya, 1954). The substance of this view is that school mathematics lessons need to involve students in doing the kind of mathematical work that resembles what mathematicians do when they work on problems (Lampert, 1990). Accordingly, classroom mathematics discussions should comprise disciplined means of reasoning that involve making conjectures, justifying them, and opening them up for others' critique and inspection. What counts as reasonable in this process is judged using available mathematical resources in the classroom community, including language, representations, and shared mathematical memory (Ball & Bass, 2003).

The Nature of Teachers' Work in Leading Classroom Mathematics Discussions

The extant research in mathematics education highlights some of the work that teachers do in managing productive classroom mathematics discussions. For instance, teachers develop classroom participation structures and intellectual and social norms that can enable students to reach sound mathematical conclusions (Ball & Bass, 2000; Lampert, 2001; McClain & Cobb, 2001). They also use public records and mathematical representations to support the development of shared mathematical knowledge (Ball & Bass, 2003; Forman & Ansell, 2002).

Research also documents some of the purposeful and disciplined *discourse routines*[1] that teachers use in managing some of the complexity inherent to this work. Discourse routines are repetitive, recognizable patterns of conversational moves in classroom mathematics discussions that involve both teacher and students. These patterns of talk are deliberately established, orchestrated, and maintained by the teacher for the purpose of bringing about student learning. Commitments to student thinking, the mathematics, and the collective as an intellectual community guide the teacher's deliberate use of these routines

(Ball & Bass, 2003). Examples of such routines include revoicing a student contribution, orienting students to each other's ideas, pressing students for explanations, connecting students' ideas, and making the structure of the mathematical discourse visible. Studies of productive classroom discussions underscore the centrality of these discourse routines in teachers' work and their effectiveness in supporting students' learning of mathematics (Chapin, O'Connor, & Anderson, 2003; Forman & Ansell, 2002; Henningsen & Stein, 1998; Rittenhouse, 1998).

The Nature of Pedagogies of Enactment in Teacher Education

Grossman and McDonald (2008) argue that teacher education needs to add pedagogies of enactment to its existing repertoire of pedagogies of investigation. Attention to the problem of enactment in teacher education presumes that equipping pre-service teachers with knowledge of the importance of discussing mathematics in the classroom or making them aware of discourse routines is not enough. Pre-service teachers also need to be able to know *how to do* things in practice and do them *interactively*.

Teaching is inherently an interactive clinical practice. Accordingly, the development of pre-service teachers' pedagogical skills requires the kind of teacher education pedagogy that is grounded both in *action* and in *reflection*. The coupling of action with reflection in teacher education pedagogy is crucial for the development of professional judgment needed in the kind of ambitious clinical practice that is responsive to learners and content (Kazemi, Lampert, & Ghousseini, 2007). Achieving these dual goals entails deliberate attention to and selection of the components of practice that can be vehicles for both enactment and reflection. It also entails the design of opportunities for what Ericsson (2002) refered to as deliberate practice that allows for "repeated experiences in which the individual can attend to the critical aspects of the situation and incrementally improve her or his performance in response to knowledge of results, feedback, or both from a teacher" (p. 368).

Grossman et al. (2009) propose what they call "approximations of practice" as affording pre-service teachers opportunities to practice elements of interactive teaching in settings of reduced complexity along with careful reflection on practice. The intervention that I describe next provides two examples of such approximations: the modeling of teaching practices and the rehearsal of teaching.

Learning to Lead Classroom Mathematics Discussions with Discourse Routines

Context

The instructional intervention was conducted in the context of a mathematics methods course for secondary pre-service teachers. Twenty three pre-service teachers were enrolled in the course. I was one of two teacher educators [TEs] who collaborated in the planning and instruction of the course. Using the instructional intervention, the TEs aimed to model the instructional practices and ways of thinking that are used in leading classroom mathematics discussions. They also aimed to ground pre-service teachers' learning about classroom mathematics discussions in the rehearsal of practices and routines of teaching. The instructional practices that were targeted included choosing a mathematical task that affords mathematical reasoning; monitoring and surfacing student thinking about the task; discussing students' thinking publicly; and managing the negotiation of a taken-as-shared knowledge. The TEs also focused on a number of discourse routines that teachers use in carrying out these practices. Examples of these routines are revoicing and orienting students to each other.

Revoicing consists of repeating, rephrasing, summarizing, elaborating, or translating someone else's speech (Chapin et al., 2003; Forman & Ansell, 2002). Repeating provides an additional opportunity for the utterance to be heard and reflected on. Elaboration allows the teacher or a classmate to reformulate the student's contribution when its author fails to make it explicit to the audience in a clear and coherent way. At the same time this move opens opportunities for the students to agree or disagree with the teacher's or classmate's characterization of the contribution. As for *orienting students to each other,* it consists of the teacher soliciting students' explanations of other students' work or their own, asking for another way to think about or solve a problem, and appealing for students' evaluations of the relevance of a student's comment (Chapin et al., 2003).

Enactment of the Instructional Intervention

The enactment of the instructional intervention started with a *model mathematics lesson,* which was followed by *reflection and discussion,* and then *guided practice.*

Model mathematics lesson. The TEs taught a model mathematics lesson that engaged the pre-service teachers in working on a mathematical problem (as students) and then discussing it as a whole class. The lesson deliberately showcased some important teaching practices and discourse routines that reflected the TE's responsiveness to the mathematics and student learning. One of the two TEs led the discussion as a mathematics teacher, while the other observed and noted aspects of teacher's practice that needed to be highlighted to the novices.

Reflection and discussion. The model mathematics lesson was then followed by a reflective discussion on the work of leading classroom mathematics discussions. The TEs deliberately labeled and elaborated certain aspects of the work of teaching that was demonstrated and opened them up for discussion. For instance, they highlighted the nature of the mathematical task chosen by the teacher during the lesson, the discourse routines used and what they afforded the TE and the students in terms of mathematical content. The following example illustrates what was said in relation to "revoicing":

> The other thing that [the teacher did] was revoicing
> [...the teacher] chose at certain points to repeat word for
> word what someone said and wrote it on the board. At
> other times, he [...] looked for someone in class and
> asked, "Can you repeat that?"

To further clarify the nature of revoicing as a discourse routine, the TE who modeled the work also reflected publicly on his use of it during the lesson, "Sometimes I tried to repeat what everyone said without being mindful about what is being said, and that had particular purposes [...] At other times, I was more deliberate in actually saying what I wanted to be said."

The pre-service teachers also participated in deliberations about the observed work of teaching and the problems of practice that emerged during the model mathematics lesson. For instance, a discussion ensued around a problem of practice that arose during the model mathematics lesson when a pre-service teacher, Norma, offered an incorrect conjecture. The TE then had invited the class to consider Norma's conjecture, which subsequently got refuted by the class. Consequently, Norma joked that she felt incompetent, hence raising concerns for some of her classmates about the affective risks entailed in the process of public refutation and validation in a mathematics discussion. The problem of practice that arose was "How does the teacher encourage disagreement while respecting students' ideas and without discouraging participation?" The TEs engaged the pre-service teachers in a careful discussion of this problem, which resulted in the consideration of alternative teaching moves that can be both attentive to important mathematical ideas and processes, and to students' sense making.

Guided practice. The discussion and reflection on the model mathematics lesson were followed by designed opportunities for the pre-service teachers to practice the use of the discourse routines. Two of these opportunities were connected in the way they were designed around a fictional classroom discussion between a teacher—Mrs. Shackleforth—and her students[2] around the following mathematical problem: *A rectangular piece of a puzzle is 4 cm x 10 cm. We want to enlarge the puzzle so that the side that measured 4 cm is now 7 cm. What should the other side measure?* The teacher lines had been erased from the dialogue, leaving lines that illustrate the different ways in which seventh graders might think of the particular problem.

In one of these two opportunities for practice, the pre-service teachers were asked to insert one or more of the discourse routines at places in the fictional discussion where they thought the teacher should have reacted to students' thinking. They were also asked to justify their choices of the routines and their placement in the discussion using annotations after each insertion. I illustrate some of this work in relation to an excerpt of the dialogue (see Figure 1), which depicts the students in Mrs. Shackleforth's class dwelling on additive rather than multiplicative reasoning.

10–Ms. Shackleforth: Here is this problem, guys. We are making a larger copy of this puzzle for the kids at the Alpha House. A rectangular piece of a puzzle is 4 cm x 10 cm. We want to enlarge the puzzle so that the side that measured 4 cm is now 7 cm. What should the other side measure?

20–Archie: Is this a trick question? It would just be 10 cm still. You just stretch it out in one direction. The other length stays the same.

30–Grace: Archie, I think you are wrong. If the side that measured 4 cm is now 7 cm then you should just add three to the other one and get 13.

40–Don: Yeah, Archie, we can't just stretch one side, because then the puzzle wouldn't fit together.

50–Archie: No, I don't mean just one side; I mean the two short sides. We stretch the two short sides and leave the two long sides alone.

60–Hank: Ermmm, I think I agree with Grace and Don more. Because like if in the first place, we had a square of 4 cm x 4 cm and we add 3 cm to all the sides, we still get a square of 7 cm x 7 cm, but if we only add 3 cm to two sides, we ain't gonna get a square no more ... and the puzzle is like kinda different, know what I mean?

Figure 1. Excerpt from the Mrs. Shackleforth Dialogue

In response to the exchange between Archie and his classmates (lines 20–50), the pre-service teachers inserted a number of discourse routines, along with justifications in the form of annotations following each insertion. Here are examples of their responses:

Brett wrote:

> Shackelforth: Ok guys, it seems that we have a bit of disagreement as to how to solve this problem. Don and Grace don't seem to agree with Archie. Can someone elaborate as to why they think that the problem must be solved in a different way? Yes, Hank? (Orienting for the purpose of negotiation).
>
> *Annotation*: Here she is trying to negotiate between the students by trying to get the students to discuss the mathematical correctness of a proposed solution.

Betty wrote:

> Shackleforth: Okay so it looks like we have a couple ideas out there. Could someone come up to the board and draw a picture of Archie's idea? Thank you, Bubba. What about Grace's idea? Thank you, Hank. (They draw pictures and label the lengths of the sides). (Orienting)
>
> *Annotation*: I hope the students will see that Archie's idea can't work because the proportions of the new shape would be too different from the original. At a minimum, the class will come to the conclusion that BOTH the length and width need to be stretched

The pre-service teachers' annotations point to their realization that there are two conjectures on the discussion floor that need to be evaluated in order to draw the students' attention to additive reasoning. This part of guided practice embodies the spirit of a pedagogy of investigation that required novices to analyze student thinking and then respond to it using a number of discourse routines.

The other practice activity that was also connected to the Mrs. Shackleforth dialogue consisted of two rehearsals of the teacher's actions. Two pre-service teachers enacted their responses to the assignment by playing the role of Mrs.

Shackleforth in a chosen segment of the discussion. The TEs intended for these rehearsals to afford the class an opportunity to witness and discuss alternative uses of the discourse routines in a common context that was familiar to all the pre-service teachers. Additionally, the rehearsals would further bring the pre-service teachers' attention to the dependence of discourse routines on contextual factors, and open up opportunities for deliberations about problems of practice and the cultivation of knowledge necessary for leading productive classroom mathematics discussions. Following these rehearsals, the pre-service teachers were allowed to revise their responses to the Mrs. Shackleforth dialogue and resubmit them for further feedback from the TEs.

The rehearsals started in a fishbowl[3] setup. A number of pre-service teachers volunteered to play the role of the students in Mrs. Shackleforth's class. They all read from the same script that was prepared by the leading pre-service teacher, who played the role of Mrs. Shackleforth. Each rehearsal lasted for 10 minutes and was followed by a discussion of the leading pre-service teacher's use of discourse routines. The TEs did not give any direct feedback to the leading pre-service teacher about her use of the routines, but rather focused the class's attention on the actions that their classmate took to support her students' learning during the discussion. For instance, in the case of one pre-service teacher, Linda, the TEs launched the discussion by asking, "What did Linda [in the role of Mrs. Shackleforth] do to make [the discussion] happen?" In response, the pre-service teachers described the different discourse moves that Linda used and the purposes they served in the discussion. For example, one pre-service teacher noted,

> "I think what [Linda] really did good [sic] was put a lot of time between comments. She did not give a comment every other student. There were some segments when there were quite a few student pieces before she would say something."

Another pre-service teacher added, "And that was good use in order to let them refute their own arguments. It gave them the opportunity to say, 'I think that one is wrong because of this.'" The TEs also participated in the noticing process by identifying particular teacher moves and highlighting their affordances for students' learning in that particular context. Beyond the noticing of productive teacher moves, additionally, this part of the rehearsal allowed the pre-service teacher who enacted Mrs. Shackleforth's role to reflect on her work. For instance, Linda shared with her classmates the dilemmas she faced in allowing students to dwell on a wrong answer. She explained, " I wanted so badly to lead to [the correct answer], and I got to really work at just making [the students] give me more wrong answers and deal with it, and it was kind of hard for me to do." In response to Linda's comment, the TEs invited the pre-service teachers to consider possible decisions Linda could have made in resolving her uneasiness with students' focus on an incorrect way of thinking. The pre-service teachers proposed several avenues of actions, which brought along opportunities to discuss appropriate judgment and relevant problems of practice.

A third practice activity was carried out in the context of the pre-service teachers' field placements, where they were asked to plan a lesson with their cooperating teachers and then lead a classroom mathematics discussion. They were expected to collect records of their practice (video, audiotape, or detailed fieldnotes taken by an observer), and then use these records to describe their work and examine reflectively their use of the discourse routines in supporting students' learning. The assignment particularly asked them to use evidence to analyze how their teaching moves may have influenced students' learning, and how they may do things differently in future lessons. Here is an example of such analysis by one pre-service teacher who found herself during the practice classroom mathematics discussion focusing on the correctness of answers rather than on the discourse process.

> I think I could have used particular discourse [routines] more effectively when I had the opportunity. For

example, at one point a student suggested squaring the numerator and denominator of a fraction. He believed the resulting fraction would be equal to the original, because "you are doing the same thing to the top and bottom." This could have been a perfect opportunity for another student to explain the error ... I think now that I could have asked students to agree or disagree with the student, or I could have asked someone to revoice what the student had suggested...

Discussion

In this chapter I illustrated how teacher education programs can focus not only on pedagogies of investigation but also on enactment. Such focus can help prospective teachers think systematically about the complexity of practice using intellectual and practical resources for use in the classroom. I have illustrated two types of pedagogies of enactment, modeling and rehearsing, which drew on intellectual and practical resources in the form of discourse routines. These two pedagogies were intertwined with investigations of practice and worked in a complementary fashion in the way the former showcased the use of particular instructional practices and made them visible to the novices, while the latter allowed the novices to experiment with them in contexts of reduced complexity.

The modeling of particular teaching practices and routines can play an important role in pre-service teachers' learning to teach in the way it showcases important aspects of the work of teaching while engaging the pre-service teachers as participants in the process (Thompson, 2006). Particularly, when modeling is coupled with the use of authentic professional learning tasks that challenge pre-service teachers' knowledge of content and engage them genuinely in the solution of mathematical problems, this pedagogy of enactment becomes more powerful in shaping novices' understanding of practice[4]. However, merely modeling particular teaching practices is not sufficient to prepare pre-service teachers to promote them in their own practice (Peressini & Knuth, 1998). Explicitly addressing the practices that are being modeled and the pedagogical thinking that is driving them is critical for what pre-service teachers learn from this pedagogy. The explicit discussion of the practices modeled by the teacher educator helps surface the elements of practice that may otherwise remain invisible to the novice. The explicitness of the modeling process also affords the development of a common language around particular teaching practices, which would enable professional education to focus its efforts on developing novices' ability to engage in such practices (Grossman & McDonald, 2008).

Rehearsal also had affordances for the pre-service teachers' development of instructional skill and the judgment necessary for using it responsively. Rehearsal offered a venue for the pre-service teachers to break out from their traditional frames of reference: it constrained them to engage in actions, which if left to their own judgment, they would not habitually use. In the case of this intervention, the discourse routines stood as the lines of a script that the novices needed to make their own in spite of their prior understandings and beliefs. In this process, the rehearsal was affording them a shift in focus to aspects of the environment they are not habitually inclined to attend to, such as the learners and their collective mathematical thinking. Such shift, over time, would prepare novices for improvisation, which requires according to Crossan (1998) "that individuals break out of their traditional frames of reference to see the environment in its full richness and complexity" (p. 595). Opportunities to lead classroom discussions in field placements were key to enriching the pre-service teachers' knowledge of the complexity of the classroom context and the way it shaped their actions as teachers.

References

Ball, D. L., & Bass, H. (2000). Making believe: The collective construction of public mathematical knowledge in the elementary classroom. In D. Phillips (Ed.), *Yearbook of the National Society for the Study of Education, Constructivism in Education* (pp. 193–224). Chicago: University of Chicago Press.

Ball, D. L., & Bass, H. (2003). Making mathematics reasonable in school. In J. Kilpatrick, W. G. Martin & D. Schifter (Eds.), *A research companion to principles and standards for school mathematics* (pp. 27–44). Reston, VA: National Council of Teachers of Mathematics.

Chapin, S. H., O'Connor, C., & Anderson, N. (2003). *Classroom discussions: Using math talk to help students learn* (Grades 1–6). Sausalito, CA: Math Solutions Publications.

Clift, R. T., & Brady, P. (2005). Research on methods courses and field experiences. In M. Cochran-Smith & K. M. Zeichner (Eds.), *Studying teacher education: The report of the AERA panel on research and teacher education* (pp. 309–424). Mahwah, NJ: Lawrence Erlbaum.

Crossan, M. M. (1998). Improvisation in action. *Organization Science, 9*(5), 593–599.

Ericsson, K. A. (2002). Attaining excellence through deliberate practice: Insights form the study of expert performance. In M. Ferrari (Ed.), *The pursuit of excellence in education* (pp. 21–55). Hillsdale, NJ: Lawrence Erlbaum.

Forman, E., & Ansell, E. (2002). Orchestrating the multiple voices and inscriptions of a mathematics classroom. *The Journal of the Learning Sciences, 11*(2&3), 251–274.

Grossman, P., Compton, C., Igra, D., Ronfeldt, M., Shahan, E., & Williamson, P. (2009). Teaching practice: A cross-professional perspective. *Teachers College Record, 111*(9).

Grossman, P., & McDonald, M. (2008). Back to the future: Directions for research in teaching and teacher education. *American Educational Research Journal, 45*, 184–205.

Hammerness, K., Darling-Hammond, L., Bransford, J., Berliner, D., Cochran-Smith, M., McDonald, M., & Zeichner, K. (2005). How teachers learn and develop. In L. Darling-Hammond, & J. Bransford (Eds.), *Preparing teachers for a changing world. What teachers should learn and be able to do* (pp. 358–388). San Francisco, CA: Jossey-Bass Educational Series.

Henningsen, M., & Stein, M.K. (1998). Mathematical tasks and student cognition: Classroom-based factors that support and inhibit high-level mathematical thinking and reasoning. *Journal for Research in Mathematics Education, 5,* 524–549.

Kazemi, E., Lampert, M., & Ghousseini, H. (2007). *Conceptualizing and using routines of practice in mathematics teaching to advance professional education.* Report to the Spencer Foundation, Chicago.

Kennedy, M. (1999). The role of pre-service teacher education. In L. Darling-Hammond,& G. Sykes (Eds.), *Teaching as the learning profession* (pp. 54–85). San Francisco, CA: Jossey-Bass.

Lampert, M. (1990). When the problem is not the question and the solution is not the answer: Mathematical knowing in teaching. *American Educational Research Journal, 27*(1), 29–63.

Lampert, M. (2001). *Teaching problems and the problems of teaching.* New Haven, CT: Yale University Press.

McClain, K., & Cobb, P. (2001). An analysis of development of sociomathematical norms in one first-grade classroom. *Journal for Research in Mathematics Education, 32*(3), 236–265.

Perissini, D. D., & Knuth, E. J. (1998). Why are you talking when you could be listening? The role of discourse and reflection in the professional development of a secondary mathematics teacher. *Teaching and Teacher Education, 14*(1), 107–125.

Polya, G. (1954). *Mathematics and plausible reasoning.* Princeton, NJ: Princeton University Press.

Rittenhouse, P. (1998). The teacher's role in mathematical conversation: Stepping in and stepping out. In M. Lampert & M. Blunk (Eds.), *Talking mathematics in school: Studies of teaching and learning* (pp. 163–189). Cambridge: Cambridge University Press.

Thompson, C. S. (2006). Powerful pedagogy: Learning from and about teaching in an elementary literacy course. *Teaching and Teacher Education, 22,* 194–204.

Endnotes

1. In conceiving these communicative actions as routines, I rely on the concept of adjacency pairs in managing verbal interaction. An adjacency pair is a sequence of two utterances which are adjacent and ordered, produced by different speakers, and typed so that each question prefers an answer, and each request a compliance (Wetherell, Taylor, & Yates, 2001).
2. The dialogue was borrowed from P. G. Herbst, who created it with his graduate students in 2004.
3. A fishbowl involves a small group of people (usually 5–8) seated in circle, having a conversation in full view of a larger group of listeners.
4. Data collected in the course of the intervention in the form of interviews with pre-service teachers gives support for this argument.

Hala Ghousseini is a post doctoral research fellow in the school of education at the University of Michigan. She works on the Learning in, from, and for Teaching Practice project and teaches in the Master of Arts and Certification Elementary Program. Previously, she taught in the secondary mathematics teacher education program at the University of Michigan and worked on the BIFOCAL project and the Michigan Mathematics and Science Teacher Leadership Collaborative, both of which support the learning of mathematics teachers in Michigan school districts. Her research focuses on studying how mathematics teaching practice can be made learnable and doable by novices, and how teacher education can be designed to combine training in skills and development of professional judgment and sensibilities.